Ernst Schering Research Foundation Workshop 9
Sex Steroids and Bone

Ernst Schering Research Foundation Workshop

Editors: Günter Stock
Ursula-F. Habenicht

Ernst Schering Research Foundation
Workshop 9

Sex Steroids and Bone

R. Ziegler, J. Pfeilschifter, M. Bräutigam,
Editors

With 35 Figures

Springer-Verlag Berlin Heidelberg GmbH

ISBN 978-3-662-03045-5   ISBN 978-3-662-03043-1 (eBook)
DOI 10.1007/978-3-662-03043-1

Originally published by Springer-Verlag Berlin Heidelberg New York in 1994
Softcover reprint of the hardcover 1st edition 1994

The use of general descriptive names, registered names, trademarks, etc. in this publica-tion does not imply, even in the absence of a specific statement, that such names are exempt from the relevant protective laws and regulations and therefore free for general use.

Product liability: The publishers cannot guarantee the accuracy of any information about dosage and application contained in this book. In every individual case the user must check such information by consulting the relevant literature.

Typesetting: Data conversion by Springer-Verlag

21/3130–5 4 3 2 1 0 – Printed on acid-free paper

Preface

Not too long ago, bone tissue was discovered to be a sex hormone-dependent organ. Similar to the development and quality of other sex hormone-dependent organs, for example, the genitalia, optimal bone mass and density and maximal function can only be achieved if the bone is exposed to sex steroids. Albright recognized the switching-off effect of menopause, leading to postmenopausal bone loss, but recent studies also throw light on the switching-on effect of puberty. Boys and girls with delayed puberty show less bone mass at the age of 20. This is of great concern since it means that these young persons might not attain their genetically allotted peak bone mass if puberty occurs too late.

The Schering Research Foundation took the opportunity to invite top notch specialists in the field to Berlin to a workshop on sex steroids and bone with the purpose of intensifying the discussions on the impact of sex hormones and associated biochemical and endocrine mechanisms on bone metabolism. The intention was to establish the state of the art on matters such as how female or male sex hormones affect the different bone cells, what the differences are, and whether we can now draw conclusions about a more differentiated therapeutic use of these hormones.

This books contains the proceedings of this workshop. The various chapters give an overall view of current knowledge on the control of bone hemostasis by sex steroids and offer new ideas on how this control could be put into practice. It was also of particular concern to integrate a general survey of in vivo experimentation and histomorphometric evaluation of bone tissue.

Abb. I. The participants of the workshop *(from left to right:* L. F. Bonewald, W. S. S. Lee, H. H. Malluche, J. C. Prior, J. Pfeilschifter, R. C. Mühlbauer, R. Ziegler, T. J. Chambers, M. Bräutigam, S. C. Manolagas, D. N. Kalu, T. J. Wronski, J. Silver)

The organizers of the workshop and editors of this volume hope that this book will contribute to a better understanding of the role of sex steroids in bone and thus pave the way for a better experimental approach in studying drug effects on bone, with the ultimate goal of improving therapy in bone diseases.

Reinhard Ziegler
Johannes Pfeilschifter
Matthias Bräutigam

Table of Contents

List of Contributors

Gideon Almogi
Minerva Center for Calcium and Bone Metabolism, Nephrology Services,
Hadassah University Hospital, Jerusalem, Israel il-91120

Lynda F. Bonewald
The University of Texas Health Science Center at San Antonio,
Department of Medicine, Division of Endocrinology and Metabolism,
7703 Floyd Curl Drive, San Antonio, TX 78284--7877, USA

Tim J. Chambers
Department of Histopathology, St George's Hospital Medical School,
London, UK

Jade Wei Mun Chow
Department of Histopathology, St George's Hospital Medical School,
London, UK

Juliet E. Compston
Dept. of Medicine, University of Cambridge Clinical School,
Addenbrooke's Hospital, Cambridge CB2 2QQ UK

Ayal Epstein
Minerva Center for Calcium and Bone Metabolism, Nephrology Services,
Hadassah University Hospital, Jerusalem, Israel il-91120

Webster S. S. Jee
Division of Radiobiology, University of Utah School of Medicine,
Salt Lake City, UT, USA

Dike. N. Kalu
Department of Physiology, University of Texas Health Science Center,
7703 Floyd Curl Drive, San Antonio, TX 78284-7756

Hua Z. Ke
Department of Cardiovascular and Metabolic Diseases,
Central Research Division, Pfizer, Inc., Eastern Point Road, Groton, CT, USA

Donald B. Kimmel
Center for Hard Tissue Research, Creighton University School of Medicine,
Omaha, NE, USA

Jennifer M Lean
Department of Histopathology, St George's Hospital Medical School,
London, UK

Mei Li
Division of Radiobiology, University of Utah School of Medicine,
Salt Lake City, UT, USA

Xiaojin J. Li
Procter and Gamble Pharmaceuticals, Miami Valley Laboratories,
Cincinnati, OH, USA

Xiaoquang G. Liang
Division of Radiobiology, University of Utah School of Medicine,
Salt Lake City, UT, USA

Baiyun Y. Lin
Division of Radiobiology, University of Utah School of Medicine,
Salt Lake City, UT, USA

Yanfei F. Ma
Division of Radiobiology, University of Utah School of Medicine,
Salt Lake City, UT, USA

Hartmut H. Malluche
Division of Nephrolog, Bone and Mineral Metabolism,
Department of Medicine, 800 Rose Street,
University of Kentucky Medical Center, Lexington, KY 40536-0084

Stavros C. Manolagas
Section of Endocrinology and Metabolism, VAMC,
and Department of Medicine, Indiana University School of Medicine,
Indianapolis, Indiana USA

Satoshi Mori
Department of Orthopaedic Surgery, University of Ryukyu Hospital,
207 Uehara, Nishihara, Okinawa, Japan 903-01

Roman C. Mühlbauer
Department of Pathophysiology, University of Berne, Murtenstrasse 35,
3010 Berne, Switzerland

Tally Naveh-Many
Minerva Center for Calcium and Bone Metabolism, Nephrology Services,
Hadassah University Hospital, Jerusalem, Israel il-91120

Jerilynn C. Prior
Departments of Medicine, Division of Endocrinology-Metabolism,
University of British Columbia, Vancouver, B. C., Canada V5Z 1 M9

Janet B. Rodgers
Division of Plastic Surgery, Department of Surgery, 800 Rose Street,
University of Kentucky Medical Center, Lexingron, KY 40536-0084

Rebecca B. Setterberg
Division of Radiobiology, University of Utah School of Medicine,
Salt Lake City, UT, USA

Justin Silver
Minerva Center for Calcium and Bone Metabolism, Nephrology Services,
Hadassah University Hospital, Jerusalem, Israel il-91120

Jonathan H Tobias
Department of Histopathology, St George's Hospital Medical School,
London, UK

Thomas J. Wronski
Department of Physiological Sciences Box 100144,
JHMHC University of Florida Gainesville, FL USA

Chiung-Fen Yen
Department of Physiological Sciences Box 100144,
JHMHC University of Florida Gainesville, FL USA

1 View of a Clinician: Sex Steroids and Osteoporosis

Juliet E. Compston

1.1 Introduction

The central role of oestrogen deficiency in the pathogenesis of postmenopausal bone loss and osteoporosis is well established. The association was first recognised by Fuller Albright in 1941 (Albright et al. 1941) and subsequently Aitken et al. (1973a) reported a reduction in metacarpal bone mass in women who had undergone oophorectomy before the age of 45 years, osteopenia developing within 3–6 years after operation. Hormone replacement therapy is the only treatment for osteoporosis in which prevention of bone loss and reduction in fracture rate in the spine, radius and hip has been definitively established. These considerations make it the therapy of choice in the prevention and treatment of osteoporosis in peri- and postmenopausal women but many questions remain unanswered.

1.2 Age-Related Bone Loss and Its Relationship to Oestrogen Status in Women

Approximately 35% and 50%, respectively, of cortical and trabecular bone are lost over a lifetime. Peak bone mass is attained during the third decade of life; subsequently there may be a period of consolidation before the onset of age-related bone loss. Some studies have demonstrated premenopausal bone loss in the spine and the femur (Riggs et al. 1986; Rodin et al. 1990), although this finding has not been universal (Aloia et al. 1985; Mazess and Barden 1991). There is, however, general agreement that rates of bone loss increase during and immediately after the menopause, particularly in the spine; thus annual rates of loss of between 1% and 6% have been reported during natural menopause (Cann et al. 1985; Hui et al. 1987) and up to 10% after oophorectomy (Genant et al. 1982). The highest rates of loss occur in trabecular bone, which has a higher surface to volume ratio than cortical bone and a correspondingly greater potential to respond to mechanical and biochemical stimuli.

1.3 Oestrogens and Bone Mass in Premenopausal Women

Oestrogen status may influence bone mass not only during and after the menopause but also premenopausally. Hypo-oestrogenic states, for example anorexia nervosa (Rigotti et al. 1984), hyperprolactinaemia (Klibanski and Greenspan 1986), amenorrhoea associated with excessive physical exercise (Drinkwater et al. 1984) and the administration of gonadotrophin-releasing hormone analogues (Matta et al. 1988) are accompanied by bone loss, and reduced bone mass has been reported in women with Turner's syndrome, which is characterised by ovarian agenesis (Shore et al. 1982). A number of studies have shown a positive association between parity and bone mass (Alderman et al. 1986), and some, but not all, studies indicate that oral contraceptive use may be associated with higher bone mass.

1.4 Effect of Oestrogen Replacement on Bone Mass and Fracture Risk

The ability of oestrogen replacement to prevent menopausal bone loss has been well documented at a number of skeletal sites, including the metacarpals, spine, radius and femur (Aitken et al. 1973b; Lindsay et al. 1980a; Ettinger et al. 1985; Stevenson et al. 1990). Although oral oestrogens were used in most of these studies, similar data also exist for transdermal and percutaneous oestrogen (Rymer et al. 1990; Stevenson et al. 1990). A reduction in fracture rate in the radius, spine and hip has also been demonstrated in women receiving hormone replacement therapy (Hutchinson et al. 1979; Weiss et al. 1980; Kiel et al. 1987). These data have been derived mainly from case-control or retrospective cohort investigations, although three prospective studies on oestrogen therapy and vertebral fracture rate have also been reported (Lindsay et al. 1980a; Riggs et al. 1982; Naessen et al. 1990). The protective effect appears to be considerable, several studies indicating an overall reduction of 50%–75% in hip fracture. However, the actual benefit may be overestimated because of intrinsic differences between women who use oestrogens during the menopause and those who do not; confounding factors due to differences in health status cannot be eliminated completely in retrospective studies. Nevertheless, the benefit is likely to be real and substantial.

1.5 Pathophysiology of Menopausal Bone Loss

Measurement of biochemical indices of bone remodelling clearly illustrate the increase in bone turnover which occurs during the menopause. Studies in oophorectomised women have shown a rapid increase in markers of bone resorption followed by a slower increase in indices of formation, changes reaching a maximum in 12–18 months (Stepan et al. 1987). Histomorphometric data in normal postmenopausal women are sparse but are consistent with increased bone turnover. Remodelling imbalance also occurs during the menopause, but it is unclear whether this is oestrogen-dependent.

It is unknown whether postmenopausal osteoporosis represents a specific disease process or is simply the result of lower peak bone mass at maturity. Several histomorphometric studies in women with postmenopausal osteoporosis have demonstrated marked heterogeneity in indices of bone turnover (Eriksen et al. 1990; Kimmel et al. 1990) which may in part reflect sequential changes, high turnover during and immediately and after the menopause being replaced in later years by low turnover. However, a remarkably consistent finding in such studies has been that of a reduction in wall width, reflecting reduced bone formation at the cellular level and suggesting the presence of a specific osteoblast defect (Arlot et al. 1984; Eriksen et al. 1990).

The increased bone turnover which occurs at the menopause has important structural consequences both in cortical and trabecular bone (Compston et al. 1989). Trabecular penetration and erosion result in reduced connectedness of the microstructure with adverse effects on bone strength and fracture risk, whilst changes at the corticoendosteal region lead to cortical thinning. The structural disruption in trabecular bone has important therapeutic implications; the ability of current therapies to restore structural integrity, once disrupted, remains unproven.

1.6 Effect of Oestrogens on Bone Remodelling and Turnover

There is strong evidence that oestrogen replacement reduces bone turnover but it is unknown whether it also has beneficial effects on remodelling balance. Remodelling balance can only be determined by histomorphometric studies and the one study in which this was assessed before and after hormone replacement therapy failed to show any significant effect (Steiniche et al. 1989); however, the relatively short treatment period of 12 months may have been insufficient to demonstrate such an effect. Reduction in bone turnover alone would be expected to lead to a transient increase in bone mass due to filling in of

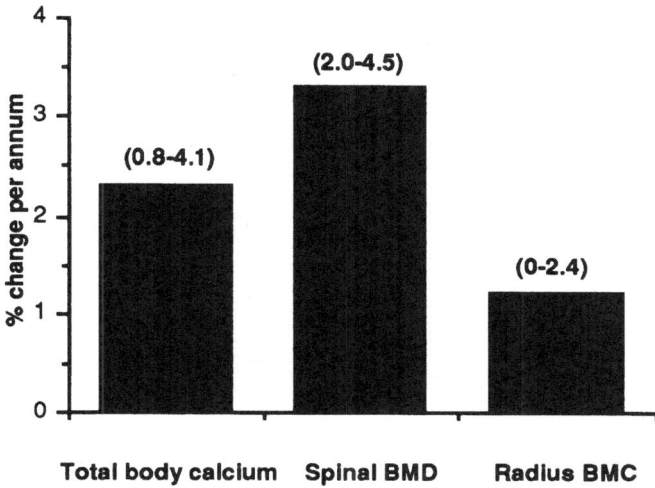

Fig. 1. Mean percentage change per annum in total body calcium, spinal bone mineral density and radial bone mineral content in 29 hysterectomised postmenopausal women treated for three years with oestradiol implants. The 95% confidence intervals of the mean change (%) for each measurement is shown in *brackets*. *BMD*, bone mineral density; *BMC*, bone mineral content

the remodelling space, and small increases after 12–18 months treatment have indeed been reported in a number of studies. Evidence that oestrogens may have true anabolic effects is provided by a recent study in which 29 hysterectomised postmenopausal women were treated with oestradiol implants for a 3-year period (Ryde et al. 1993). In this study total body calcium was assessed by prompt γ neutron activation analysis, spinal trabecular bone density by quantitative computed tomography and mid-radius bone mineral content by single photon absorptiometry. The results are shown in Table 1 and Fig. 1; total body and regional bone mass increased progressively over the study period, with mean annual increases of 2.4% in total body calcium, 3.3% in spinal bone mineral density and 1.2% in radial bone mineral content. These significant increases in bone mass over a period of 3 years thus strongly suggest that oestrogens increase bone mass in menopausal women; however, histomorphometric data are required to establish the mechanisms by which this increase occurs.

Table 1. Mean and standard error of the mean (SEM) values for normalised total body calcium, spinal trabecular bone mineral density, and radial bone mineral content

Measurement				
	First	Second	Third	Fourth
Normalised TBCa (cts)				
Mean	0.0128	0.0132	0.0136	0.0137
SEM	0.0003	0.0004	0.0004	0.0004
n	29	29	27	23
QCT (mg/ml)				
Mean	152.3	163.1	162.6	165.8
SEM	5.5	5.3	5.5	5.7
n	29	29	27	23
BMC (g/cm)				
Mean	0.884	0.883	0.889	0.900
SEM	0.026	0.023	0.024	0.029
n	29	29	27	23

TBCA, total body calcium; QCT, quantitative computed tomography; BMC, bone mineral content

1.7 Assessment of Fracture Risk

The development of noninvasive techniques for assessment of bone mass at clinically relevant sites enables detection of low bone mass before fracture occurs and provides a means by which fracture risk may be predicted. Bone mass is a major determinant of fracture risk, although other factors, particularly trauma, also contribute; in addition, changes in trabecular microstructure and in bone matrix and mineral composition are likely to play a role. A number of prospective studies have investigated the relationship between bone mass and fracture risk in postmenopausal women (Gardsell et al. 1989; Hui et al. 1988; Cummings et al. 1990); overall, these studies indicate an increasing fracture risk with decreasing bone mass, a decrease of 1 SD in bone mass being accompanied by a relative risk for fracture of 1.5–3; the predictive value of bone mass measurement is greatest if performed at the site of potential fracture, at least for the hip (Cummings et al. 1993). Most of these studies have been performed in women in the seventh and eighth decades of life and there are relatively few prospective data relating bone mass at the menopause to future fracture risk. In this context the issue of whether a subgroup of menopausal women are "fast losers" is important, since a single measurement of bone mass at the menopause would not distinguish these from those with slower rates of loss.

The predictive value of clinical and historical risk factors has been investigated in several studies. Overall, these are only weakly predictive for bone mass or fracture risk but strong risk factors are of clinical value when making decisions about bone densitometry and/or treatment; these include a past history of fragility fracture, premature menopause and long-term treatment with high doses of corticosteroids (Compston 1992).

The issue of screening for osteoporosis using bone mass measurements has been hotly debated but most agree that such an approach cannot be justified at present because of the paucity of prospective data relating bone mass at the menopause to future fracture risk and the inability accurately to assess cost/effectiveness of screening (Melton et al. 1988). The alternative to screening is to select women for densitometry on the basis of risk; in practice, decisions about selection for densitometry are heavily influenced by the resources available. Current recommendations for bone densitometry include the presence of radiological osteopenia and/or vertebral deformity, past history of fragility fracture, prolonged secondary amenorrhoea and long-term corticosteroid therapy; since hormone replacement is now prescribed routinely for women with a premature menopause (< 40 years), bone densitometry is unnecessary unless specific contraindications to hormone replacement exist or the demonstration of low bone mass is required before treatment will be accepted (Johnston et al. 1989).

1.8 Indications for Long-Term Hormone Replacement Therapy

Although there are some advocates of life-long hormone replacement therapy for all women, these are in a minority. A case for such an approach might be made if the cardioprotective effects of combined hormone replacement therapy were shown to be equal to (or greater than)

those of unopposed oestrogen, but at present it is unknown how progestogens affect oestrogen-induced cardiovascular benefits. Furthermore, the probable increase in breast cancer risk would deter many
women from long-term therapy, particularly if no clear indications
were present. The alternative to the above policy is to select women for
treatment on the basis of fracture risk, assessed by bone densitometry
and/or strong clinical risk factors such as premature menopause, secondary amenorrhoea, fragility fracture and corticosteroid therapy.

Densitometric criteria for the prevention and treatment of osteoporosis have not been clearly defined. Because bone mass in the majority
of the elderly female population is near or below the notional fracture
threshold, bone density values in postmenopausal women are usually
related to those obtained in healthy premenopausal women; a bone
density more than two standard deviations below the mean reference
value for this population would be regarded by most as a firm indica-
tion for treatment but whether, as some have advocated (Johnston et al.
1989), all those with values below one standard deviation of the reference premenopausal mean value should be included is more debatable.
Rigid criteria for treatment require the development of tested screening
protocols which provide treatment guidelines based on bone density
and fracture risk.

1.9 Duration of Treatment

The optimal duration of treatment with hormone replacement therapy
in the prevention or treatment of osteoporosis is unresolved. Several
case-control studies suggest that the benefits associated with current
oestrogen use may not be maintained some years after stopping therapy, but there is no direct evidence to confirm or refute this contention
and the tendency of such studies to be biased by a high proportion of
long-term and current users makes conclusions about duration of the
protective effect speculative. Figure 2 shows data from two prospective studies in which the age adjusted risk of hip fracture was calculated according to time since last taking hormone replacement therapy
(Kiel et al. 1987; Paganini-Hill et al. 1981). Whilst these data may indicate that the protective effect becomes weaker after therapy is
stopped, they do not take into account the duration of oestrogen use.

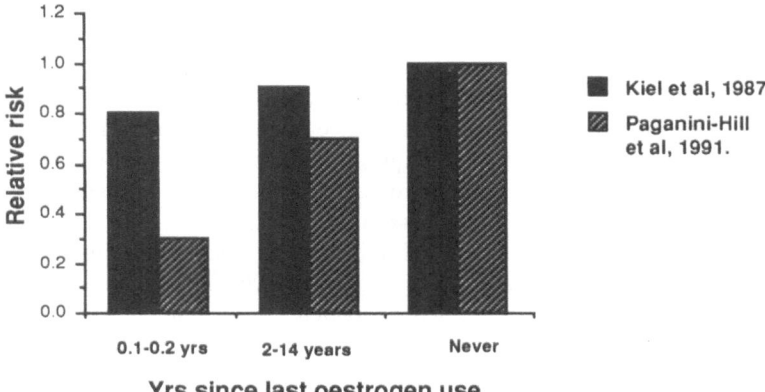

Fig. 2. Age-adjusted risk of hip fracture since last taking hormone replacement therapy. Data obtained from Kiel et al. 1987 *(black bars)* and Paganini-Hill et al. 1981 *(hatched bars)*

There is similar uncertainty about the progress of bone loss after hormone replacement therapy is stopped. In a study of oophorectomised women, Lindsay et al. (1978) showed that although oestrogen treatment for 8 years prevented significant bone loss, metacarpal bone mass in patients treated only for the first 4 years after oophorectomy did not differ significantly at 8 years from that in women treated with placebo, due to an acceleration in the rate of bone loss after treatment in the group treated for only 4 years. In contrast, a cross-over study in postmenopausal women suggested that the rate of metacarpal bone loss after cessation of active therapy was similar to that of untreated women (Christiansen et al. 1980). There are no comparable studies in which changes in bone mass in the spine and femur have been examined after treatment is stopped.

Until these uncertainties are resolved, decisions about duration of treatment must be based on assessment of the relative risks and benefits of hormone replacement therapy; as discussed below, this presents further problems. The present concensus is that hormone replacement therapy should be prescribed for 5–10 years in order to strike a balance between the undoubted skeletal benefits and possible adverse effects on the breast. However, this length of treatment may be

insufficient to offer significant protection against hip fracture, which
has its highest incidence in the ninth decade of life, some 20–30 years
after hormone replacement therapy would be stopped. Prospective
studies of hormone use and hip fracture pose formidable problems because of the long time span involved; even with vertebral fractures,
10–15 years follow-up may be required. If the assumption is made that
the protective effect of hormone replacement therapy is predominantly
due to preservation of bone mass, investigation of changes in bone
mass after cessation of therapy would indicate whether lasting benefit
was obtained from treatment for 5 years; however, such data are sparse
and more information is needed. In patients with a premature menopause, particularly if this occurs before the age of 40, the duration of
treatment required may be longer than 10 years, but, once again, de-
finitive evidence is lacking.

1.10 Timing of Therapy

There is general agreement that hormone replacement therapy should be
started early in the menopause since bone mass at that time will be
higher than if treatment is postponed for 5, 10 or more years. However,
in view of the lack of data relating to changes in bone mass after treatment is stopped and the possibility that any benefits gained from 5 years
of treatment may be lost by the time hip fracture risk is greatest, this
practice might be questioned. Hormone replacement therapy prevents
age-related bone loss even if started many years after the menopause but
data on fracture risk in these groups are not available (Quigley et al.
1987). Whether bone mass is higher in the eighth and ninth decades of
life in women treated for a finite period early in the menopause is higher
than in those treated for the same period ten or more years after the
menopause is unknown; "rebound" bone loss, if it occurs, may be more
severe in the former group when natural rates of bone loss are higher.
However, the acceptability of hormone replacement therapy is likely to be
much greater in women in the early stages of the menopause because of
the high incidence of vasomotor and other menopausal symptoms; when
started at a later stage, side effects of treatment are more common and
the prospect of withdrawal bleeding may be unacceptable.

1.11 Unopposed Versus Opposed Therapy

The demonstration that oestrogen therapy was associated with an increased risk of endometrial cancer and the subsequent evidence that this could be eliminated by the addition of a progestogen for 10–12 days of each cycle has led to the widespread prescription of opposed therapy to all women with an intact uterus. The skeletal benefits of opposed therapy appear to be similar to those of treatment with oestrogen alone; indeed, there is evidence that progestogens may independently reduce menopausal bone loss (Gallagher et al. 1991). Nevertheless, the use of progestogens in hormone replacement formulations creates some problems. Side effects such as headaches, abdominal bloating, nausea and psychological complaints are more common with combined preparations and there are theoretical grounds for believing that progestogens may adversely affect the cardiovascular benefits associated with long-term menopausal oestrogen use. In patients with a past or present history of cardiovascular disease a case can therefore be made for the use of unopposed oestrogen even in nonhysterectomised women, provided that regular endometrial monitoring is performed. In view of the low grade of malignancy and excellent prognosis of endometrial cancers occurring in oestrogen-treated women and the significant reduction in cardiovascular morbidity and mortality associated with oestrogen use, unopposed therapy might be considered justifiable for all women but medicolegal considerations make this course of action problematic unless specific contraindications to opposed treatment exist.

1.12 Risks and Benefits of Long-Term Hormone Replacement Therapy

Evaluation of the cost-effectiveness of long-term hormone replacement therapy depends critically upon accurate information about both its risks and benefits. Present information is derived almost exclusively from unopposed oestrogen preparations, which offer substantial cardiovascular benefits but appear to be associated with some increase in the relative risk of breast cancer. The central question of whether and how progestogens affect these issues is unresolved.

   The reduction in ischaemic heart disease reported in menopausal oestrogen users is of the order of 50% (Bush et al. 1987; Barrett-Connor and Bush 1991; Wolf et al. 1991), although the magnitude of this effect may have been overestimated in observational studies because of their inability to correct for confounding factors related to the health status of oestrogen users and nonusers. The mechanism by which this protective effect occurs is poorly understood, but is likely to be related at least in part to higher high density lipoprotein (HDL) and lower low density lipoprotein (LDL) cholesterol levels in oestrogen users. Oestrogen-induced changes in glucose metabolism, coagulation and fibrinolysis may also contribute.

There is now considerable evidence that oestrogen therapy for longer than 10 years increases the risk of breast cancer in postmenopausal women (Bergkvist et al. 1989; Colditz et al. 1990). The magnitude of increase in risk appears to be small and is not associated with an increased mortality. High-dose preparations are associated with a greater risk but it is unknown whether risk is influenced by the type of oestrogen (natural or synthetic) or the route of administration.

In terms of mortality, the potential cardiovascular effects of hormone replacement therapy overshadow those both on the skeleton and breast, and if combined therapy was shown to be equally cardioprotective to unopposed therapy a case could be made for hormone replacement therapy in all women as prophylaxis against heart disease. Anxiety about breast cancer would, however, make many women reluctant to embark upon such a course, particularly if positive indications for therapy were lacking. The question of whether progestogens affect breast cancer risk is highly relevant since if, as suggested by one study, they were protective against breast cancer, willingness to take and compliance with hormone replacement therapy would be greatly increased.

1.13 Formulations of Hormone Replacement Therapy

The rapid scientific and clinical advances in the fields of bone physiology and osteoporosis have not yet been matched by improvement in formulations of hormone replacement therapy. Although most preparations use natural rather than synthetic oestrogens, horse urine remains the most common source of oestrogen used and the majority of com-

bined formulations still use androgenic rather than progestogenic progestogens.

Transdermal administration of oestrogen with or without progestogen avoids the hepatic first pass effect and provides a more physiological serum hormonal profile than oral administration, in which high
peak levels are required and unfavourable oestrodiol/oestrone ratios
produced. The effects of transdermal preparations on bone mass are
comparable to those achieved with oral oestrogens although there are
as yet no data on fracture risk. Whether avoidance of the hepatic first
pass effect modifies any of the risks and benefits of long-term hormone
replacement therapy has not been established. Percutaneous oestrogens
(implants) also avoid first pass hepatic metabolism; however, their insertion involves a minor surgical procedure and the plasma oestradiol
levels achieved are extremely variable, leading to severe side effects in
some women. Tachyphylaxis has also been reported, symptoms recurring after decreasing periods of time after implantation despite extremely high plasma oestradiol concentrations.

1.14 Future Directions

Improvement of existing hormone replacement formulations is being
explored in a number of different ways, including the use of nonan-
drogenic progestogens and improved methods for transdermal administration of oestrogens and progestogens. Perhaps the most exciting
area is the development of more selective oestrogen agonists and anta-
gonists, which may ultimately obviate the need for a progestogen and
eliminate the increase in risk of breast cancer. The example of tamoxifen has demonstrated the potential for both oestrogenic and anti-oestrogenic properties within a single compound, anti-oestrogenic effects
in the breast being associated with oestrogenic effects in the endometrium and probably also the skeleton. Figure 3 shows the tissue-based
bone formation rate in a group of women treated for breast cancer with
tamoxifen compared to age-matched women with breast cancer not
treated with tamoxifen; the significantly lower value in the tamoxifen-
treated group indicates oestrogenic effects on bone and is in keeping
with densitometric data from other studies (Wright et al. 1993).

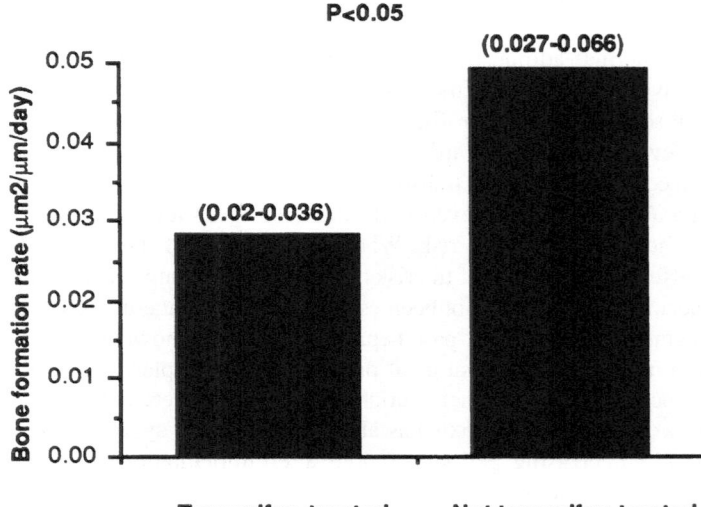

Fig. 3. Tissue-based bone formation rate in 19 women on long-term tamoxifen therapy for breast cancer compared with values in 15 age-matched women with breast cancer not treated with tamoxifen. Data are shown as the median with the interquartile range *(brackets)*. Data from Wright et al. (1993)

Approaches which reduce the frequency of or eliminate the need for vaginal bleeding in women with an intact uterus are also being explored. The synthetic steroid tibolone combines weak oestrogenic, androgenic and progestogenic properties; it is effective in the treatment of vasomotor symptoms associated with the menopause and does not cause significant endometrial hyperplasia. It has been shown to prevent menopausal bone loss in the spine, femur and radius in relatively short-term studies but no fracture data are available (Lindsay et al. 1980b; Rymer et al. 1990; Geusens et al. 1991). Avoidance of vaginal bleeding may also be achieved by continuous administration of a low dose of progestogen with an oestrogen throughout the 28-day cycle; alternatively the length of the cycle may be increased to reduce the frequency of vaginal bleeding.

Other future prospects include the synthesis of nonsteroidal compounds with oestrogenic activity, the incorporation of bone-seeking compounds such as bisphosphonates into the oestrogen molecular structure to target activity to the skeleton and the addition of anabolic agents, for example, bone morphogenetic proteins and insulin-like growth factors, to hormone replacement preparations. The ideal agent would combine suppressive effects on bone turnover with anabolic effects and would retain the cardiovascular benefits of oestrogens whilst avoiding adverse effects on the breast and endometrium. Finally, restoration of trabecular microstructure in patients with established osteoporosis is a major target for future drug development.

References

Aitken JM, Hart DM, Anderson JB, Lindsay R, Smith DA, Speirs CF (1973a) Osteoporosis after oophorectomy for non-malignant disease in premenopausal women. Br Med J 2:325–328

Aitken JM, Hart DM, Lindsay R (1973b) Oestrogen replacement therapy for prevention of osteoporosis after oophorectomy. Br Med J 3:515–518

Albright F, Smith PH, Richardson AM (1941) Postmenopausal osteoporosis. JAMA 116:2465–2474

Alderman BW, Weiss NS, Daling JR, Ure CL, Ballard JH (1986) Reproductive history and postmenopausal risk of hip and forearm fracture. Am J Epidemiol 124:262–267

Aloia JF, Vaswani A, Ellis K, Yuen K, Cohn SH (1985) A model for involutional bone loss. J Lab Clin Med 106:630–637

Arlot M, Edouard Meunier PJ, Neer RM, Reeve J (1984) Impaired osteoblast function in osteoporosis: comparison between calcium balance and dynamic histomorphometry. Br Med J 289:517–520

Barrett-Connor E, Bush TL (1991) Estrogen and coronary heart disease in women. JAMA 265:1861–1867

Bergkvist L, Adami HO, Persson I, Hoover R, Schairer C (1989) The risk of breast cancer after oestrogen and oestrogen-progestin replacement. N Engl J Med 321:293–297

Bush TL, Barrett-Connor E, Cowan LD et al (1987) Cardiovascular mortality and non-contraceptive use of estrogen in women: results from the Lipid Research Clinics Program Follow-up Study. Circulation 75:1102–1109

Cann CE, Genant HK, Kolb FO, Ettinger B (1985) Quantitative computed tomography for prediction of vertebral fracture risk. Bone 6:1–7

Christiansen C, Christensen MS, McNair PL, Hagen C, Stocklund KE, Transbol I (1980). Prevention of early postmenopausal bone loss: conducted 2-year study in 315 normal females. Eur J Clin Invest 10:273–279

Colditz G A, Stampfer MJ, Willett WC, Hennekens CH, Rosner B, Speizer FE (1990). Prospective study of estrogen replacement therapy and risk of breast cancer in postmenopausal women. JAMA 264:2648–2653

Compston JE (1992) Risk factors for osteoporosis. Clin Endocrinol 36:223–224

Compston JE, Mellish RWE, Croucher PI, Newcombe R, Garrahan NJ (1989) Structural mechanisms of trabecular bone loss in man. Bone Miner 6:339–350

Cummings SR, Black DM, Nevitt MC et al (1990) Appendicular bone density and age predict hip fractures in women. JAMA 263:665–668

Cummings SR, Black DM, Nevitt MC et al (1993) Bone density at various sites for predicition of hip fractures. Lancet 341:72–75

Drinkwater BL, Nilson K, Chesnut CH, Bremmer WJ, Shainholtz S, Southworth MB (1984) Bone mineral content of amenorrhoeic and eumenorrhoeic athletes. N Engl J Med 311:277–281

Eriksen EF, Hodgson SF, Eastell R, Cedel SL, O'Fallon WM, Riggs BL (1990) Cancellous bone remodelling in type I (postmenopausal) osteoporosis: quantitative assessment of rates of formation, resorption and bone loss at tissue and cellular levels. J Bone Miner Res 5:311–319

Ettinger B, Genant HK, Cann CE (1985) Long-term oestrogen replacement therapy prevents bone loss and fractures. Ann Intern Med 102:319–324

Gallagher JC, Kable WT, Goldgar D (1991) Effect of progestin therapy on cortical and trabecular bone: comparison with estrogen. Am J Med 90:171–178

Gardsell P, Johnell O, Nilsson BE, Nilsson JA (1989) The predictive value of fracture, disease and falling tendency for fragility fractures in women. Calcif Tissue Int 45:327–330

Genant HK, Cann CE, Ettinger B, Gilbert SG (1982) Quantitative computed tomography of vertebral spongiosa: a sensitive method for detecting early bone loss after oophorectomy. Ann Intern Med 97:699–705

Geusens P, Dequecker J, Gielen J, Schot LPC (1991) Non-linear increase of vertebral density by a synthetic steroid (Org OD 14) in established osteoporosis. Maturitas 13:155–162

Hui SL, Slemenda CW, Johnston CC, Appledorn CR (1987) Effects of age and menopause on vertebral bone density. Bone Miner 2:141–146

Hui SL, Slemenda CW, Johnston CC (1988) Age and bone mass as predictors of fracture in a prospective study. J Clin Invest 81:1804–1809

Hutchinson A, Polansky SM, Feinstein AR (1979) Post-menopausal oestrogens protect against fractures of the hip and distal radius: a case-control study. Lancet ii:705–709

Johnston CC, Melton LJ, Lindsay R, Eddy DM (!989) Clinical indications for bone mass measurements. J. Bone Miner Res 4 [Suppl 2] 1–28

Kiel DP, Felson DT, Anderson JJ, Wilson PWF, Miskowitz MA (1987) Hip fracture and the use of oestrogens in postmenopausal women: the Framingham Study. N Engl J Med 317:1169–1174

Kimmel DB, Recker R R, Gallagher JC, Vaswani AS, Aloia JF (1990) A comparison of iliac bone histomorphometric data in postmenopausal osteoporotic and normal subjects. Bone Miner 11:217–235

Klibanski A, Greenspan SL (1986) Increase in bone mass after treatment of hyperprolactinaemic amenorrhoea. N Engl J Med 315:542–546

Lindsay R, Hart DM, MacLean A, Clark AC, Kraszewski A, Garwood J (1978) Bone response to termination of estrogen treatment. Lancet i:1325–1327

Lindsay R, Hart DM, Forrest C, Baird C (1980a) Prevention of spinal osteoporosis in oophorectomised women. Lancet ii:1151–1153

Lindsay R, Hart DM, Kraszewski A (1980b) Prospective double-blind trial of synthetic steroid (Org OD14) for preventing postmenopausal osteoporosis. Br Med J 280:1207–1209

Matta WH, Shaw R W, Hesp R, Evans R (1988) Reversible trabecular bone density loss following induced hypo-oestrogenism with the GnRH analogue buserelin in premenopausal women. Clin Endocrinol 29:45–51

Mazess RB, Barden HS (1991) Bone density in premenopausal women: effects of age, dietary intake, physical activity, smoking, and birth-control pills. Am J Clin Nutr 53:132–142

Melton LJ, Kan SH, Wahner HW, Riggs BL (1988) Lifetime fracture risk: an approach to hip fracture risk assessment based on bone mineral density and age. J Clin Epidemiol 41:985–994

Naessen T, Persson I, Adami HO, Bergstrom R, Bergkvist L (1990) Hormone replacement therapy and risk for first hip fracture. Ann Intern Med 113:95–103

Paganini-Hill A, Ross RK, Gerkins VR, Henderson BE, Arthur M, Mack TM (1981) Menopausal estrogen therapy and hip fractures. Ann Intern Med 95:28–31

Quigley MET, Martin PL, Burnier AM, Brooks P (1987) Oestrogen therapy arrests bone loss in elderly women. Am J Obstet Gynecol 156:1516–1523

Riggs BL, Seeman E, Hodgson SF, Taves DR, O'Fallon WM (1982) Effect of the fluoride/calcium regimen on vertebral fracture occurrence in postmenopausal osteoporosis. N Engl J Med 306:446–450

Riggs BL, Wahner HW, Melton LJ, Richelson LS, Judd HL, Offord KP (1986) Rates of bone loss in the appendicular and axial skeletons of women. Evidence of substantial vertebral bone loss before menopause. J Clin Invest 77:1487–1491

Rigotti NA, Nussbaum SR, Herzog DB, Neer RM (1984) Osteoporosis in women with anorexia nervosa. N Engl J Med 311:1601–1606

Rodin A, Murby B, Smith MA et al (1990) Premenopausal bone loss in the lumbar spine and neck of femur: a study of 225 Caucasian women. Bone 11:1–5

Ryde SJS, Bowen-Simpkins K, Bowen-Simpkins P, Evans WD, Morgan WD, Compston JE (1990) The effect of oestradiol implants on regional and total bone mass; a three year longitudinal study. Clin Endocrinol (in press)

Rymer J, Chapman M, Fogelman I (1990) OD 14: Bone protection with no endometrial stimulation. In: Christiansen C, Overgaard K (eds) Osteoporosis 1990: Third International Symposium. ApS, Copenhagen, Denmark, pp 1071–1074

Shore RM, Chesney RW, Mazess RB, Rose PG, Bergman GJ (1982) Skeletal demineralisation in Turner's syndrome. Calcif Tissue Int 34:519–522

Steiniche T, Hasling C, Charles P, Eriksen EF, Mosekilde L, Melsen F (1989) A randomised study on the effects of oestrogen/gestagen or high dose oral calcium on trabecular bone remodelling in postmenopausal osteoporosis. Bone 10:313–320

Stepan JJ, Pospichal J, Presl J, Pacovsky V (1987) Bone loss and biochemical indices of bone remodelling in surgically induced postmenopausal women. Bone 8:279–284

Stevenson JC, Cust MP, Gangar KF, Hillard TC, Lees B, Whitehead MI (1990) Effects of transdermal versus oral hormone replacement therapy on bone density in spine and proximal femur in postmenopausal women. Lancet 335:265–269

Weiss NS, Ure CL, Ballard JH, Williams AR, Daling JR (1980) Decreased risk of fractures of the hip and lower forearm with postmenopausal use of oestrogens. N Engl J Med 303:1195–1198

Wolf PH, Madans JH, Finucane FF, Higgins M, Kleinman JC (1991) Reduction of cardiovascular disease-related mortality among postmenopausal women who use hormones: evidence from a national cohort. Am J Obstet Gynecol 164:489–494

Wright CDP, Mansell RE, Gazet JC, Compston JE (1993) Effect of long term tamoxifen treatment on bone turnover in women with breast cancer. Br Med J 306:429–430

2 The Anabolic Action of Estrogen on Rat Bone

Tim J. Chambers, Jade Wei Mun Chow, Jennifer M. Lean, and Jonathan H. Tobias

2.1 Background

Postmenopausal osteoporosis is due to estrogen deficiency: similar bone loss develops after ovariectomy, in amenorrheic women, and in other estrogen-deficient states (Schlechter et al. 1983; Cann et al. 1984; Koppelman et al. 1984; Rigotti et al. 1984) and can be prevented by estrogen administration (Lindsay et al. 1980; Ettinger et al. 1985). The bone loss is associated with an increase in bone resorption, accompanied by an increase in bone formation that is insufficient to compensate for the increased resorption (Christiansen et al. 1982; Steiniche et al. 1989). The increased bone formation is caused by the increased resorption rather than by estrogen deficiency itself, since inhibition of bone resorption by agents such as calcitonin (CT) and bisphosphonate also suppress the increased bone formation. Thus, estrogen acts in humans to maintain bone mass through inhibition of bone resorption.

Estrogen appears to play a similar role in the rat: ovariectomy or suppression of 17β-estradiol production by busurelin induces rapid bone loss, which is associated with increased resorption and increased bone formation (Wronski et al. 1988; Goulding and Gold 1990). E_2 replacement, CT and bisphosphonate all suppress bone resorption and bone formation also falls (Wronski et al. 1989, 1991a,b; Hayashi et al. 1989).

The mechanism by which E_2 suppresses bone resorption is uncertain. It appears to be a direct effect on bone (Takano-Yamamoto and Rodan 1990) and inhibits osteoclasts either directly (Oursler et al. 1991) or via osteoblasts (Tobias and Chambers 1991). We also, unexpectedly, found a direct stimulation of osteoclastic activity in vitro by E_2, at high concentrations, similar to those that might occur in pregnancy. It seemed possible that this might function to release a restraint on bone resorption during late pregnancy, to facilitate transfer of calcium to the fetus (Chambers and Tobias 1990). To see if the same occurred in vivo, we administered high doses of E_2 to rats.

2.2 Experimental Evidence

We found no evidence for suppression of resorption parameters by high doses of E_2 (Tobias et al. 1991). This is unlike the clear decrease observed when lower dosages are administered and supports the possibility that high concentrations of E_2 might indeed enable an increase, compared to lower concentrations, in bone resorption. However, we also noted, to our surprise, that E_2 stimulated bone formation. We subsequently found that concentrations of E_2 close to or within the physiological range were also anabolic (Chow et al. 1992a,b) and that bone showed a sensitivity to the effects of E_2 in the same dose-response range as other recognized targets of E_2 (see Fig. 1).

In these experiments, increased bone formation was measured as an increase in fluorochrome-labeled surfaces. A theoretical objection could be that since E_2 acts to inhibit resorption, increased fluorochrome label might represent suppression of label resorption rather than increased formation. Against this, 3-amino-1-hydroxypropylidene-1-bisphosphonate (AHPrBP), which is a more potent inhibitor of bone resorption than is E_2, reduced rather than increased fluorochrome per-

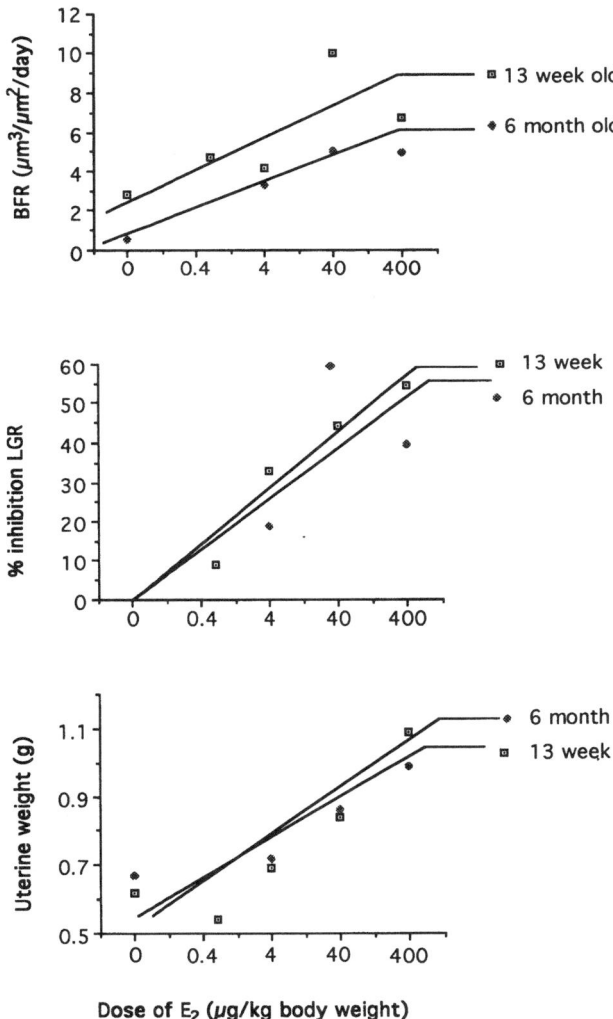

Fig. 1. Relationship between the dose of E$_2$ administered and the bone formation rate *(BFR)*, percent inhibition of longitudinal growth rate *(LGR)*, and uterine weight in 13-week-old and 6-month-old animals. (From Chow et al. 1992b)

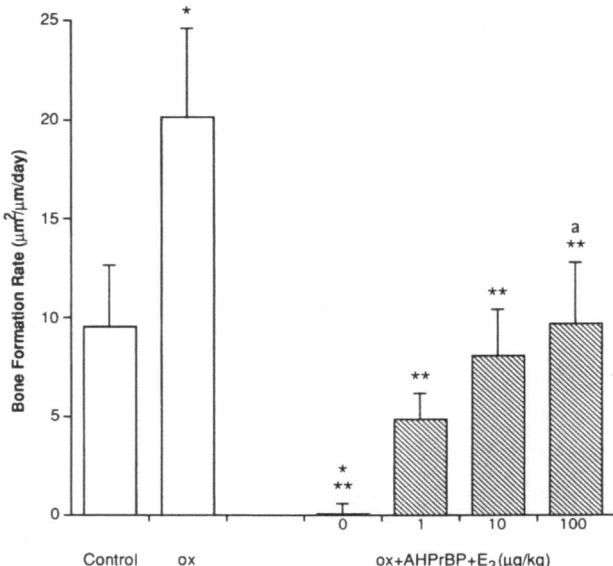

Fig. 2. Effect of ovariectomy *(OX)*, 3-amino-1-hydroxypropylidene-1-bisphos-
phonate *(AHPrBP)*, and E_2 on bone formation rate at the proximal tibial metaphysis. Results are expressed as mean ± SEM. *Significantly ($p < 0.05$) different from control; **significantly different ($p < 0.05$) from OX; [a]significantly
different ($p < 0.05$) from OX + AHPrBP. (From Chow et al. 1992a)

imeters (Chow et al. 1992a; see Fig. 2). Nor could the increased label
be an artifact due to suppression of longitudinal growth by E_2-in-
creased fluorochrome labeling was observed even at doses of E_2 too
small to affect longitudinal growth. We have, moreover, observed a
similar increase in fluorochrome labeling in animals inspected 24 h
after labeling, too short a time for significant resorption of such recently deposited mineral to have occurred (submitted). Finally, a similar increase is seen if bone forming surface is assessed as osteoblast
surface or as mineralizing surface in the scanning electron microscope
(in preparation).

Another theoretical objection to the notion that these observations
represent a true anabolic effect would be that in a system such as the
rat metaphysis, where resorption and formation are coupled, sup-

pression of resorption might accelerate the onset of the coupled, subsequent phase of bone formation. If this were the case, a similar increase in bone formation would be expected after bisphosphonate administration, but none was observed (Chow et al. 1992a; see Fig. 2). Moreover, the bone formation induced by E_2 does not occur on recently-resorbed surfaces (see below). Our conclusion from many experiments is that E_2 exerts not only an anticatabolic but also an anabolic effect on rat cancellous bone.

The anabolic effect is, however, not sustained: it appears to be detectable by 4 days, to reach a peak at approximately 2 weeks, and to decline by 3 weeks. This later decline explains why others have not noted an anabolic effect, since previous experiments have always measured bone formation in the rat after a longer duration of administration; an anabolic effect in humans might have been similarly missed.

One possible explanation for the transient nature of the anabolic response might be that E_2 acts to establish a new set point for bone volume (Turner 1991) and that once this has been achieved, the bone formation rate returns to normal. To test this, we rendered rats osteopenic by ovariectomy: when osteopenia had developed, 13 weeks after ovariectomy, Estrogen was administered for 8 weeks. We noted a subnormal bone formation rate in E_2-treated rats, despite a subnormal bone volume (Abe et al. 1993).

A second possible explanation for the transient nature of the anabolic response is that some other ovarian principle is required, in addition to E_2, for maintenance of bone formation. There is, though, no change in the response of bone to E_2 in the presence of a progesterone inhibitor. We (unpublished) and others (Kalu et al. 1991) have found no effect of progesterone or progesterone inhibitor (Abe et al. 1992b) on bone in the rat, and the anabolic effect of E_2 is transient even in intact rats (Tobias et al. 1993).

A third possibility is that, in view of the coupling between bone resorption and bone formation, the anabolic action of E_2 may be suppressed by its own antiresorptive action. In the osteopenic rats above (Abe et al. 1993), residual trabeculae were still capable of extensive bone-forming activity, as judged from the high bone formation rate of ovariectomized but otherwise untreated animals. Yet despite this, and despite a subnormal bone volume, bone formation rate was reduced rather than increased by E_2. Indeed, the effect of E_2 was indistinguishable from that of the other inhibitors of bone resorption used in that experiment (CT and AHPrBP), in which inhibition of bone resorption was associated with suppression of bone formation. For E_2, either tachyphylaxis occurs to the anabolic more rapidly than to the antiresorptive action; or the antiresorptive action itself curtails expression of the anabolic effect.

The available evidence suggests that the anabolic effect of E_2 is curtailed by its antiresorptive activity. We noted that, in the presence of AHPrBP, a more potent inhibitor of bone resorption than is E_2, E_2 induced a smaller absolute increase in bone formation than when given alone (compare Chow et al. 1992a,b). We therefore formally tested the effect of AHPrBP on the action of E_2, and found that AHPrBP significantly suppressed the anabolic action of E_2 (Abe et al. 1992a). This effect of AHPrBP was unlikely to be a direct, toxic effect on bone formation, since bone formation was unaffected at modeling sites such as the diaphysis: it appears to be an effect on bone formation mediated through resorption suppression, similar to the action on bone formation of CT, and of E_2 itself.

2.3 Concluding Remarks

We conclude from our experiments that E_2 has, in addition to its antiresorptive action, an anabolic effect on rat trabecular bone, in the physiological range. This conclusion is consistent with an increasing body of direct and indirect evidence: ovariectomy in beagle dogs has been found to reduce mean wall thickness (Malluche et al. 1986); administration of relatively low doses of E_2 to rabbits causes osteoid accumulation (Whitson 1972); in mice, pharmacological doses of E_2 cause intense accumulation of trabecular bone (Urist et al. 1950; Morse et al. 1974; Simmons 1963, 1966); E_2 stimulates medullary bone in birds (Pfeiffer and Gardner 1938); in humans, at puberty, an increase in bone mass correlates with estrogen levels (Gilsanz et al. 1991). The in vitro evidence suggests that E_2 is anabolic (Ernst et al. 1989; Komm et al. 1988), and we have found that if a supraphysiological dose of E_2 is discontinued after prolonged administration to intact rats, there is a decline in the bone formation rate; and if bone formation rate is assessed before increased resorption entrains increased formation, it is reduced after ovariectomy. The rapid bone loss, in humans and the rat, after ovariectomy is generally attributed to increased turnover. However, other conditions in which bone turnover is increased through increased resorption (e.g., hyperparathyroidism) are not necessarily associated with bone loss (Silverberg et al. 1989). Similarly, suppression of resorption is not associated with a continued increase in bone volume. Thus, increased bone turnover in estrogen-deficient states does not itself cause bone loss, but facilitates expression of an underlying, unrelated drive towards reduced bone volume (Parfitt 1979). Our data suggest that loss of the anabolic action of estrogen might represent part of this drive and account for the discrepancy between the resorption and formation of bone that causes bone loss in estrogen-deficient states. Thus, to the extent that increased bone resorption does not itself of necessity lead to reduced bone volume, the loss of the anabolic action of estrogen might be of greater significance in the pathogenesis of the osteopenia caused by estrogen deficiency than the increased resorption.

It seems perverse that an agent which, in the rat at least, is capable of inducing bone formation should impair its own ability to do so, through suppression of bone resorption. It may be significant that bone is normally exposed to intermittent rather than sustained E_2 levels. We have compared the effects of intermittent with continuous E_2 treatment in the rat and found a greater increase in bone volume with intermittent treatment (submitted). It is tempting to speculate that E_2 supports a natural ADFR (activate resorption; depress resorption; free of drug interference; repeat) regime (Frost 1980), such that variations in hormonal levels, which are known to occur during the ovarian cycle in humans and the rat, initially (when low) facilitate resorption, and then (when high) enhance the subsequent, coupled, formative phase.

It seems plausible that a hormone which plays a crucial role in the determination of bone volume should be able to increase bone mass both through preventing bone loss and increasing bone formation. Although much of the experimental evidence for this derives from the rat, it is worth noting that for those observations that have been made in humans under similar circumstances to those in the rat, the human data conforms to rat observations. Thus, in humans, as in the rat, estrogen deficiency increases resorption and formation, and CT, bisphosphonate and E_2 suppress both; and all prevent but do not restore bone loss due to E_2 deficiency. The acute effects of E_2 administration in humans have not been assessed.

References

Abe T, Chow JWM, Lean JM, Chambers TJ (1992a) The anabolic action of 17β-estradiol (E_2) on rat trabecular bone is suppressed by (3-amino-1-hydroxypropylidene)-1-bisphosphonate (AHPrBP). Bone and Mineral 19: 21–29

Abe T, Chow JWM, Lean JM, Chambers TJ (1992b) The progesterone antagonist, RU 486, does not affect basal or estrogen-stimulated cancellous bone formation in the rat. Bone Miner 19: 225–233

Abe T, Chow JWM, Lean JM, Chambers TJ (1993) Estrogen does not restore bone lost after ovariectomy in the rat. J Bone Miner Res 8:831–838

Cann CE, Martin MC, Genant HK, Jaffe RB (1984) Decreased spinal mineral content in amenorrheic women. JAMA 251:626–629

Chambers TJ, Tobias JH (1990) Role of estrogens in the regulation of bone resorption. In: Nordin BEC (ed) Osteoporosis: contributions to modern management. Parthenon, Carnforth, pp 21–30

Chow J, Tobias JH, Colston KW, Chambers TJ (1992a) Estrogen maintains trabecular bone volume in rats not only by suppression of bone resorption but also by stimulation of bone formation. J Clin Invest 89:74–78

Chow JWM, Lean JM, Chambers TJ (1992b) 17β-estradiol stimulates cancellous bone formation in female rats. Endocrinology 130: 3025–3032

Christiansen C, Christiansen MS, Larsen NE, Transbol I (1982) Pathophysiological mechanism of estrogen effect on bone metabolism: dose-response relationship in early postmenopausal women. J Clin Endocrinol Metab 55: 1124–1130

Ernst M, Heath JK, Rodan GA (1989) Estradiol effects on proliferation, messenger ribonucleic acid for collagen and insulin-like growth factor-I, and parathyroid hormone-stimulated adenylate cyclase activity in osteoblastic cells from calvariae and long bones. Endocrinology 125:825–833

Ettinger B, Genant HK, Cann CE (1985) Long-term estrogen replacement therapy prevents bone loss and fractures. Ann Int Med 102:319–324

Frost HM (1980) The ADFR concept and monitoring it. In: Jee WSS and Parfitt AM (eds) Bone Histomorphometry. Third International Workshop, Sun Valley. Metab Bone Dis Rel Res 2 [Suppl]:317–321

Gilsanz V, Rose TF, Mora S, Costin G, Goodman WG (1991) Changes in vertebral bone density in black girls and white girls during childhood and puberty. N Engl J Med 325:1597–1600

Goulding A, Gold E (1990) Buserelin-mediated osteoporosis: effects of restoring estrogen on bone resorption and whole body calcium content in the rat. Calcif Tissue Int 46:14–19

Hayashi T, Yamamuro T, Okumura H, Kasai R, Tada K (1989) Effect of (Asu[1,7])-eel calcitonin on the prevention of osteoporosis induced by combination of immobilization and ovariectomy in the rat. Bone 10:25–28

Kalu DN, Salerno E, Liu CC, Echon R, Ray M, Garza-Zapata M, Hollis BW (1991) A comparative study of the actions of tamoxifen, estrogen and progesterone in the ovariectomized rat. Bone Miner 15:109–124

Komm BS, Terpening CM, Benz DJ, Graeme KA, Gallegos A, Korc M, Greene GL, O'Malley BW, Haussler MR (1988) Estrogen binding, receptor mRNA and biologic response in osteoblast-like osteosarcoma cells. Science 241:81–84

Koppelman MCS, Kurtz DW, Morrish KA, Bou E, Susser JK, Shapiro JR, Loriaux DL (1984) Vertebral body bone mineral content in hyperprolactinemic women. J Clin Endocrinol Metab 59:1050–1053

Lindsay R, Hart DM, Forrest C, Baird C (1980) Prevention of spinal osteoporosis in oophorectomized women. Lancet ii:1151–1153

Malluche HH, Faugere M-C, Rush M, Friedler RM (1986) Osteoblastic insufficiency is responsible for maintenance of osteopenia after loss of ovarian function in experimental Beagle dogs. Endocrinology 119:2649–2654

Morse BS, Giuliani D, Soremekun M, Difino S, Giuliani R (1974) Adaptation of hemopoietic tissue resulting from estrone-induced osteosclerosis in mice. Cell Tissue Kinetics 7:113–123

Oursler MJ, Osdoby P, Pyfferoen J, Riggs BL, Spelsberg TC (1991) Avian osteoclasts as estrogen target cells. Proc Natl Acad Sci USA 88:6613–6117

Parfitt AM (1979) Quantum concept of bone remodeling and turnover: implications for the pathogenesis of osteoporosis. Calcif Tissue Int 28:1–5

Pfeiffer CA, Gardner WU (1938) Skeletal changes and blood serum calcium level in pigeons receiving estrogens. Endocrinology 23:485–491

Rigotti NA, Nussbaum SR, Herzog DB, Neer RM (1984) Osteoporosis in women with anorexia nervosa. N Engl J Med 311:1601–1606

Schlechter JA, Sherman R, Martin R (1983) Bone density in amenorrheic women with and without hyperprolactinemia. J Clin Endocrinol Metab 56:1120–1123

Silverberg SJ, Shane E, De La Cruz L, Dempster DW, Feldman F, Seldin D, Jacobs TP, Siris ES, Cafferty M, Parisien MV, Lindsay R, Clemens TL, Bilezikian JP (1989) Skeletal disease in primary hyperparathyroidism. J Bone Miner Res 4:283–291

Simmons DJ (1963) Cellular changes in the bones of mice as studied with tritiated thymidine and the effects of estrogen. Clin Orthop 26:176–189

Simmons DJ (1966) Collagen formation and endochondral ossification in estrogen-treated mice. Proc Soc Exp Biol Med 121:1165–1168

Steiniche T, Hasling C, Charles P, Eriksen EF, Mosekilde L, Melsen F (1989) A randomised study on the effects of estrogen/gestagen or high dose oral calcium on trabecular bone remodeling in postmenopausal osteoporosis. Bone 10:313–320

Takano-Yamamoto T, Rodan GA (1990) Direct effects of 17β-estradiol on trabecular bone in ovariectomised rats. Proc Natl Acad Sci USA 87:2172–2176

Tobias JH, Chambers TJ (1991) Effect of sex hormones on bone resorption by rat osteoclasts. Acta Endocrinol 124:121–127

Tobias JH, Chow J, Colston KW, Chambers TJ (1991) High concentrations of 17β-estradiol stimulate trabecular bone formation in adult female rats. Endocrinology 128:408–412

Tobias JH, Chow JWM, Chambers TJ (1993) Increase in bone volume in the rat following intermittent but not continuous 17β-estradiol. Proceedings of the 4th International Symposium on Osteoporosis, Hong Kong, March/April 1993, pp 156–157

Turner CH (1991) Homeostatic control of bone structure: an application of feedback theory. Bone 12:203–217

Urist MR, Budy AM, McLean FC (1950) Endosteal bone formation in estrogen-treated mice. J Bone Joint Surg 32A:143–162

Whitson SW (1972) Estrogen-induced osteoid formation in the osteon of mature female rabbits. Anat Rec 173:417–436

Wronski TJ, Cintron M, Dann LM (1988) Temporal relationship between bone loss and increased bone turnover in ovariectomized rats. Calcif Tissue Int 43:179–183

Wronski TJ, Dann LM, Scott KS, Crooke LR (1989) Endocrine and pharmacological suppressors of bone turnover protect against osteopenia in ovariectomized rats. Endocrinology 125:810–816

Wronski TJ, Yen CF, Burton KW, Mehta RC, Newman PS, Soltis EE, DeLuca PP (1991a) Skeletal effects of calcitonin on ovariectomized rats. Endocrinology 129:2246–2250

Wronski TJ, Yen CF, Scott KS (1991b) Estrogen and diphosphonate treatment provide long-term protection against osteopenia in ovariectomized rats. J Bone Miner Res 6:387–394

3 Progesterone and Its Role in Bone Remodelling

Jerilynn C. Prior

3.1 Introduction

Preliminary evidence that progesterone can act as a bone-trophic hormone was initially collected in 1990 (Prior 1990). Since then a number of further reports are available which modify and extend the concepts presented in that paper. In particular, more work has recently been published on growth factors and progesterone in bone metabolism, on changes in bone markers during the luteal phase or with progestin therapy, and in both animal and human models of progesterone treated, reproductively mature individuals immediately after oophorectomy.

As increasing clinical data are collected, a more detailed model of progesterone action on bone is needed that will take into account the activation frequency and the rate of bone resorption. A hypothesis is advanced that progesterone's potential bone-trophic effects can only be manifest if bone resorption rates are low or if activation frequency is slow. Therefore, progesterone treatment can increase bone density but only in bones that are in a state of low bone turnover. When bone turnover is high, progesterone must be given with antiresorption treatments, such as oestrogens, bisphosphonates or calcitonin, to produce a net beneficial effect on bone density.

The purpose of this review is to present new evidence relating bone metabolism to progesterone and its synthetic and clinically useful analogues, progestins. In addition, this paper will present the hypothesis that progesterone's net effect on bone density is strongly influenced not only by its intrinsic actions which promote bone formation, but also by the rate of bone remodelling. This postulate will be illustrated by presenting rates of change in bone density in women treated immediately after or several years following premenopausal oophorectomy. Finally, this review will present pilot clinical studies of cyclic medroxyprogesterone treatment in amenorrhoeic young women and continuous progestin treatment in osteopenic postmenopausal women with contraindications to oestrogen treatment.

3.2 Osteoblast Progesterone Receptors and Receptor-Mediated Actions

The initial in vitro studies of progesterone-mediated actions on osteoblast-like cells in culture implied that progesterone competes with glucocorticoids for a glucocorticoid osteoblast receptor (Manolagas and Anderson 1978). Since 1990, studies have been published from several centres using different cell lines and different progestins that all show progesterone and its analogues to have receptor-mediated osteoblast actions (Masuda and Guo 1991; Demarest et al. 1991; Tremollieres et al. 1991, 1992; Tertinegg and Heersche 1992). Several of these studies indicate that progesterone leads to increased tritiated thymidine incorporation by osteoblasts (Demarest et al. 1991; Tremollieres et al. 1991, 1992; Schevens et al. 1992), suggesting the proliferation of new osteoblast cells.

Medroxyprogesterone, dydrogesterone, cyproterone, norethindrone and promogestone (but not norgesterol which, like norethindrone and promogestone, is a 19-nor-testosterone derived progestin) also show evidence that increased DNA synthesis (increased thymidine incorporation) is stimulated when they are added to osteoblast cells in culture (Demarest et al. 1991; Tremollieres et al. 1991; Schevens et al. 1992). In addition, progesterone, norethindrone and medroxyprogesterone all appear to stimulate osteoblastic differentiation as evidenced by increased alkaline phosphatase release into the cell culture medium (Demarest et al. 1991; Schevens et al. 1992). In the most detailed study currently available, both progesterone and promogestone have dose-dependent stimulatory effects on the human osteoblast-like cell TE85, increasing the cell numbers by 125%–150% over doses from 0.01 to 100 nmol/l, therefore spanning the physiological range of doses (Tremollieres et al. 1992). Finally, bone nodule formation in rat calvarial cell cultures was shown to be stimulated in a dose-dependent fashion by progesterone in vitro (Tertinegg and Heersche 1992).

It is possible that there is more than one osteoblast progesterone receptor since a recent report suggests that antiprogesterone (RU 486), anti-oestrogen (tamoxifen) and anti-androgen (flutamide) compounds did not interfere with the progestin-stimulated proliferation of mouse osteoblasts in culture (Demarest et al. 1992). Although these are early studies and most have not yet been published, the congruity of the evidence from several laboratories with different techniques and osteoblast cell lines and culture conditions tends to lend credence to the concept that progesterone and progestins act directly on osteoblasts through specific receptors.

3.3 Growth Factors–Interactions with Insulin, Progesterone and Bone Cells

The discussion of growth factors in bone is introduced in this review because several lines of evidence link the osteoblast cellular actions of progesterone to various growth factors. There is increasing evidence that growth factors of various kinds are active in bone turnover (Mohan and Baylink 1991). Although both insulin-like growth factors I and II (IGF-I and -II) are found in bone tissue, in most species, except in the rodent, IGF-II is believed to be the important bone growth factor (Mohan and Baylink 1991). IGF-II has significant chemical homology with, and may be identical to, the previously described "skeletal growth factor" (Mohan and Baylink 1991). IGF-I is understood to be the same as somatomedin C which modulates growth hormone action in tissues (Mohan and Baylink 1991). Insulin, as well as both IGF-I and IGF-II have been shown to promote the dose-dependent growth of osteoblasts in cell culture (Masuda and Guo 1991; Tremollieres et al. 1992; Lempert et al. 1992).

Insulin is known to cause a dose-dependent increase in osteoblast cell growth in culture (Masuda and Guo 1991; Felsenfeld et al. 1992). In the same human osteoblast-like cells (HOS TE85), IGF-I showed a similar dose-response proliferation (Masuda and Guo 1991). When progesterone in a concentration of 10 nM was added to the culture medium of the HOS cells, there was an early (24 h) and marked increase in release of a unique IGF binding protein (35 kDa) while estradiol, testosterone, 1,25-dihydroxyvitamin D$_3$, growth hormone and retinoic acid were negative at the same concentrations (Masuda and Guo 1991). In a different laboratory, similar in vitro studies using the same TE85 cell line as well as normal human bone cells characterized as osteoblasts, progesterone and promogestone (a 19-nor-progesterone compound) in 5 nM concentrations led to increases in IGF-II but not IGF-I secretion (Tremollieres et al. 1991). These investigators also found that the mRNA for IGF-II was increased when TE85 cells were exposed to progesterone and promogestone (Tremollieres et al. 1991). In further studies, progesterone plus IGF-II induced an increase in cell proliferation over that caused by either agent alone (Lempert et al. 1992). In addition, progesterone appeared to stimulate increased mRNA levels for IGF-II, IGF-I, IGF binding protein 5, type 1 and type

2 receptors while the decreasing levels of mRNA for the inhibitory IGF binding protein 4 (Lempert et al. 1992; Tremollieres et al. 1992).

The links between growth factors in bone and progesterone have also been shown in clinical as well as in vitro studies. When large doses of medroxyprogesterone (500 mg/d i.m.) were given to women with advanced breast cancer, the serum levels of IGF-I doubled ($n = 7$, $p = 0.02$) and the levels of IGF binding protein 1 decreased by 75% ($n = 7$, $p = 0.05$) although levels of IGF-II were unchanged (Reed et al. 1992). However, similar women with advanced breast cancer treated with the anti-oestrogen tamoxifen showed the opposite changes: significant decreases in the serum levels of IGF-I ($n = 9$, $p = 0.05$) and increases in the IGF-binding protein 1 ($n = 9$, $p = 0.001$) (Reed et al. 1992). In another human study, ovulatory women were found to have significant increases in the level of somatomedin C (another name for IGF-I) during the luteal phase when progesterone levels were increased 1400 times over their follicular phase baseline (Nielsen et al. 1990).

In summary, although the physiological significance of the relationships between progesterone, progestins, growth factors and bone metabolism remain unclear, consistent recurrent reports of their associations continue to appear. In human studies and in cell culture experiments progesterone treatment leads to increases in osteoblast cell growth and differentiation as well as to increased production of bone growth factors and changes in the concentrations of growth factor binding proteins.

3.4 Markers of Bone Remodelling

Direct indices of bone turnover and changes in bone metabolism, such as with histomorphometric bone biopsy studies, are inconvenient, invasive and expensive. Therefore markers of bone metabolism that can be measured in serum and in urine become important. At the time of the initial review of progesterone and bone (Prior 1990) numerous studies showed that calcium and hydroxyproline excretions decreased during treatment with progestins and a few reports indicated that total alkaline phosphatase stayed the same or increased but few data were available concerning the newer indicators of bone formation (such as plasma levels of osteocalcin, procollagen peptide, and bone-specific alkaline phosphatase) or of bone resorption (such as plasma levels of tartrate-resistant acid phosphatase and the excretion of pyridinoline cross-links or n-telopeptide). This section will summarize the new data on bone markers and progesterone's bone actions.

During one ovulatory menstrual cycle from each of eight women, significant increases in osteocalcin and bone-specific alkaline phosphatase levels (as well as increases in somatomedin C, as mentioned previously) were found during the luteal phase (Nielsen et al. 1990). Although these authors found no correlations between luteal phase increases in osteocalcin, bone-specific alkaline phosphatase and somatomedin C levels with progesterone levels, it is likely that time series analysis or evaluations using log-transformed data would be necessary for the association to be shown.

In animal and human progesterone treatment studies fairly consistent results indicate that markers of bone formation are increased or not suppressed, and that markers for bone resorption, if elevated, are not suppressed to normal and are unchanged if low. Kalu et al. (1991) recently showed that after 40 weeks of progesterone treatment in oophorectomized mature rats serum osteocalcin levels were significantly increased over the levels of sham-operated or placebo-treated oophorectomized animals (both $p < 0.001$). It is of interest to note that in the same study, rats treated with oestrogen alone or oestrogen plus progesterone showed suppressed levels of serum osteocalcin.

Two human treatment studies measured osteocalcin and bone alkaline phosphatase levels prospectively after premenopausal oophorectomy and both show these bone formation markers increase associated with progestin treatment. Stepan and colleagues (1989) began treating 22 women who were 3 months after premenopausal oophorectomy – 11 were assigned to norethisterone (5 mg/day) and 11 to transdermal oestrogen (50 μg/day). Serum and urine bone markers were prospectively measured during the progestin (6 months of treatment) and the oestrogen (4 months of treatment) trial. Markers of resorption, serum levels of tartrate-resistant acid phosphatase and hydroxyproline and calcium excretions, which were markedly elevated initially (see subsequent section on oophorectomy), were decreased significantly by norethisterone treatment but the mean values did not return to normal (Stepan et al. 1989). However, bone alkaline phosphatase levels tended to increase more gradually and osteocalcin levels which were initially increased remained elevated. In contrast, during oestrogen treatment skeletal alkaline phosphatase decreased and osteocalcin levels were reduced to baseline values, suggesting that oestrogen decreases the increased bone formation rates induced by oophorectomy as well as suppressing resorption (Stepan et al. 1989).

The second study of bone markers during clinical progesterone treatment examined changes in bone turnover during 5 weeks in steroid-treated asthmatic men given 200 mg. i.m. of depo-medroxyprogesterone and 12 healthy age- and sex-matched controls (Grecu et al. 1991). At baseline the controls and asthmatic men had similar levels of vitamin D hormones, IGF, and total and bone alkaline phosphatase but osteocalcin and calcitonin levels were decreased ($p < 0.001$). In response to treatment, osteocalcin levels increased by 110% by the first week and declined but remained elevated by the fifth week. Calcitonin levels increased by 90% at the second week and declined but remained 40% elevated at week 5. Bone-specific alkaline phosphatase levels were gradually increased to reach a significant 32% elevation by the third week, which remained throughout the study (Grecu et al. 1991). This study, as the previous ones, indicates that markers of bone formation are increased in response to progesterone treatments.

3.5 Oophorectomy Model–Rapid Bone Turnover Influences Progesterone Effects

The effects of gonadal steroid hormones on bone density and bone balance could be ideally tested in an animal in which the endogenous hormone levels are low and therefore a known dose of the particular steroid can be added and the results monitored. The oophorectomized mature animal would seem to fulfil all of the requirements for an ideal model. However, bone tissue has the unique characteristic of "coupled remodelling" to consider. It is now clear that major changes in bone remodelling occur acutely with certain hormonal changes. These abrupt changes include the onset of glucocorticoid treatment (LoCascio et al. 1990), the onset of hypothalamic amenorrhoea (Biller et al. 1991), hypogonadism induced acutely by gonadotrophin agonist therapy (Survey and Judd 1992), and, most dramatically and classically, with pre-

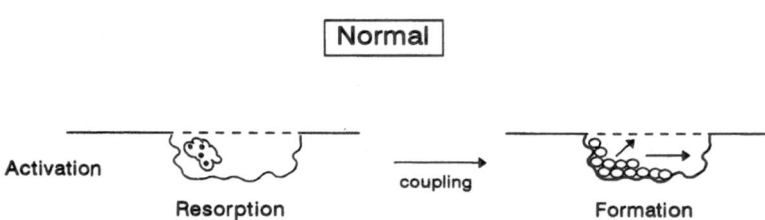

Fig. 1. The normal bone remodelling cycle with appropriate activation frequency and normal bone resorption coupled to equal bone formation

menopausal oophorectomy (Ohta et al. 1992). Therefore, before we can consider the action of progesterone on bone in the oophorectomized woman or animal, we need to reconsider the potential effect of the dramatic increase in remodelling that occurs after oophorectomy.

Bone is composed of millions of "bone remodelling units", discrete areas of trabecular bone which each act in a unified manner to facilitate the renewal of bone cells and structure (Parfitt 1984). Bone remodelling allows bone that had been laid down years before to be resorbed and replaced with new, and presumably, stronger bone. Although careful time-controlled histomorphometric studies can tell us about cellular activities, the biochemical, biomechanical, growth factor and hormonal stimulae of these actions and their interactions are still being described.

It is useful here to review normal bone remodelling, how it is altered by premenopausal oophorectomy or the onset of amenorrhoea, and subsequently the net effect of progesterone treatment acutely and after the rate of remodelling has slowed. As shown in Fig. 1, a given bone remodelling unit is first activated by largely unknown processes. After activation, osteoclasts, bone marrow-derived multinucleated giant cells, increase in both number and activity. Bone resorption occurs as the osteoclasts dig out a portion of bone to a given depth (approximately 37 μm on average) over about 12 days (Reeve 1987). Then resorption slows and over the next 27 days as-yet-unknown signals couple the resorption phase of bone remodelling to the next stage – this is called reversal. The bone formation phase is much slower, lasting about 94 days, during which osteoblast cells, derived from

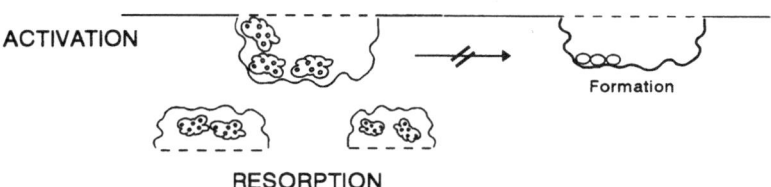

Fig. 2. In contrast to Fig. 1, immediately after an abrupt change in gonadal steroid hormone levels (such as premenopausal oophorectomy, the onset of amenorrhoea, or treatment with gonadotropin-releasing hormone agonists) there is an increase in bone activation associated with marked increases in bone resorption. However, this diagram shows that resorption and formation are no longer coupled and formation is low, leading to a net bone loss

mesenchymal tissues, begin to increase in number and subsequently in activity (Reeve 1987). Osteoblasts utilize calcium, phosphate and matrix proteins to lay down new bone. As the new bone reaches some new prescribed level, but commonly the level of the previous old bone, that particular bone remodelling unit becomes quiescent and the same process begins again in another unit. In the time course of bone remodelling in a given bone multicellular unit the resorption phase is rapid and therefore dominates if activation frequency is increased. The whole process of a given remodelling cycle takes about 4 months in humans.

Sudden and profound decreases in the gonadal hormones, oestrogen and progesterone, such as occur immediately after premenopausal oophorectomy, with the onset of amenorrhoea, and with gonadotrophin agonist treatment are associated with dramatic increases in the rate of bone remodelling (Karambolova et al. 1986; Barengolts et al. 1990; Biller et al. 1991; Ohta et al. 1992, Reeve 1987) (Fig. 2).

Modelling the changes based on Stepan's hydroxyproline data (Stepan et al. 1987), Reeve states that "A doubling of the mean rates of bone resorption seems to occur rapidly after ovariectomy" (Reeve

1987). First, multiple bone remodelling units are activated at the same time. Secondly, osteoclasts are increased both in number and in their resorptive activity. Resorption cavities are therefore not only more prevalent but deeper. Thirdly, and very importantly, the formation phase of remodelling is no longer tightly coupled to resorption. Formation rates may be higher than normal, normal, or markedly decreased (conflicting reports are available) but whatever formation rates are found are, according to the concepts of coupling, inadequate to equal the rate of resorption (Mazzuoli et al. 1990). The net effect is a rapid phase of cortical (Meema and Meema 1968; Horsman et al. 1977) as well as an even more dramatic decrease in trabecular bone, with losses of 5%–10% in 1 year, as has been shown to occur after oophorectomy in trabecular spinal bone density measured by quantitative computed tomography (QCT) (Cann et al. 1980).

The therapeutic implications of these postulates about the time course of bone remodelling after oophorectomy are three: (1) these states of rapid bone turnover and loss should be prevented if clinically possible, (2) treatment needs to begin immediately following surgery, the use of gonadotropin-releasing hormone (GnRH) agonists or the onset of amenorrhoea, and (3) antiresorption agents with or without progestins would more successfully prevent bone loss. One would also expect that agents such as progestins, which are weak antiresorbers, would not prevent initial bone loss even though formation parameters are increased. Given that low bone turnover states with markedly decreased rates of bone turnover as well as low bone formation subsequently develop (Reeve 1987; LoCascio et al. 1990), progesterone and progestins would seem to be appropriate treatment several years after oophorectomy.

Studies of oestrogen and progesterone treatment in the oophorectomized adult rat tend to confirm these ideas (Kalu et al. 1991). After 40 weeks of treatment with supraphysiological doses of oestrogen (serum levels were twice those of sham-operated animals) or with physiological levels of progesterone, the proximal tibial bone volume was increased in the oophorectomized estradiol (E_2)-treated animals but as low as in the untreated oophorectomized rats in those treated with progesterone (Kalu et al. 1991). However, histomorphometry showed that the proximal tibial cancellous bone apposition rate was normal in the progesterone-treated animals compared to low bone for-

Progesterone Treatment

Amenorrhoea of <1 year's duration

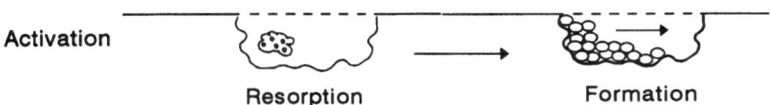

Fig. 3. If progesterone treatment is given acutely after the abrupt change in hormones and in the face of increased bone turnover, although formation will be increased, the net effect will still be loss of bone

mation rates in the E_2-treated rats. Furthermore, osteocalcin levels, which reflect bone formation or turnover, were markedly increased by progesterone treatment but suppressed with oestrogen treatment (Kalu et al. 1991). Progesterone treatment of oophorectomized 10-week-old rats resulted in a net loss of bone compared to sham-operated animals but a higher bone density than in the untreated oophorectomized animals (Barbagallo et al. 1989). Those animals that received supraphysiological doses of oestrogen had bone densities that equalled sham-operated animals (Barbagallo et al. 1989).

Studies of oophorectomy in the aged rat model also provide evidence that progesterone increases bone formation (Barengolts et al. 1990). In this model the treatment was for 20 weeks and showed that progesterone treatment (in physiological doses that reproduced a normal progesterone level) was associated with decreased osteoclast numbers and percent of osteoclast surfaces as well as increasing measures of bone formation (mineral appositional rate and percent of double-labelled surfaces) (Barengolts et al. 1990). Furthermore, this study showed, for the first time, that progesterone treatment was associated with preservation of mechanical strength by the three-point bending test (Barengolts et al. 1990). Barengolts and colleagues, in a study as yet only published in abstract, that oestrogen and progesterone have equivalent effects. Progesterone treatment in the rat, in contrast to the human treated immediately after oophorectomy, appears to decrease

Table 1. Net spinal bone density changes in response to 10 mg/day medroxy-progesterone related to number of years after premenopausal oophorectomy

	Age	Years (POO)	Height (cm)	Weight (kg)	Ll–L4 Begin	Ll–L4 End	Change per year	%Change per year
Early treatment								
	49	0	159.5	55.0	1.005	0.895	−0.110	−10.94
	48	0	151.5	75.4	1.567	1.483	−0.084	−5.36
	42	0	159.5	72.2	1.466	1.425	−0.041	−2.80
Mean	46.3	0.0	156.8	67.5	1.346	1.267	−0.078	−6.4
SD	± 3.8		± 4.6	± 10.9	± 0.30	± 0.32	± 0.03	± 4.2
Late treatment								
	42	7	183.0	65.0	0.951	0.980	0.029	3.05
	41	11	177.0	90.0	0.863	0.893	0.030	3.48
Mean	41.5	9	180.0	77.5	0.907	0.936	0.029	3.3
SD	± 0.7	± 2.8	± 4.2	± 17.7	± 0.6	± 0.6	± 0.0	± 0.3

POO, postophororectomy

the increased rate of bone resorption to normal and therefore a net positive bone balance can be achieved.

Bisphosphonates, which, like oestrogen, function predominantly to decrease bone resorption, show effects similar to oestrogen in clinical studies in women following oophorectomy (Smith et al. 1989). A hypothesis related to women treated after oophorectomy is that the increased resorption rate induced by surgery is not appropriately counterbalanced by an increased formation rate which also occurs acutely after oophorectomy (Ohta et al. 1992). The additive effect on formation produced by progesterone treatment does not appear to be sufficient to prevent a net loss of bone in the first year after surgery (Fig. 3).

Controlled prospective studies in women are not available to confirm or refute this notion. However, some clinical data are available. We have recently observed that the rate of change in bone density per year in women who were untreated for many years following premenopausal oophorectomy was significantly positive while the rate of change in those treated acutely after surgery was negative (Table 1). Preliminary observations of three women ages 42 to 49 treated with medroxyprogesterone 10 mg/day starting 4 days following premeno-

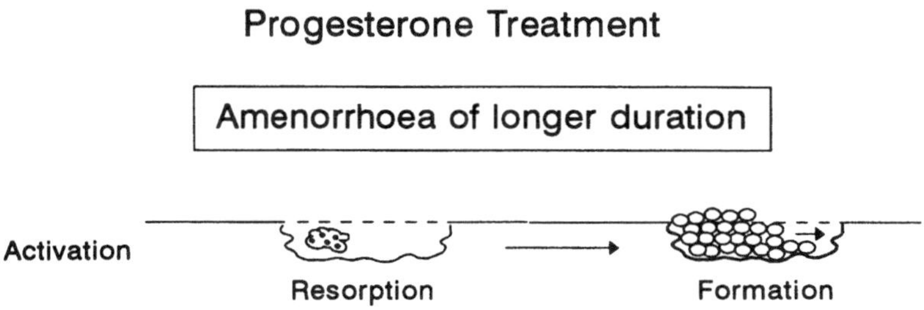

Progesterone Treatment

Amenorrhoea of longer duration

Activation

Resorption   Formation

Fig. 4. If progesterone treatment is given to an oophorectomized woman many years after her premenopausal surgery, bone turnover will be minimal. Because remodelling is less, progesterone can have a major effect on bone formation with a net gain in bone density

pausal surgery compared with two women ages 41 and 42 treated 11 and 7 years after oophorectomy with the same progestin and dose show this contrast. In the early treatment group bone density change in 1 year by dual energy X-ray absorptiometry (DXA) showed a loss of 6.4% while the two women treated years after oophorectomy showed increases of 3.1% and 3.5% consistent with the model as shown in Fig. 4.

Considering the acute ovarian hormonal suppression produced by gonadotrophin agonists (GnRHa) for the treatment of endometriosis to be similar to premenopausal oophorectomy (which it is in vasomotor symptoms, oestrogen levels and vaginal atrophy; Survey and Judd 1992), bone change when "add-back" progestin treatment is simultaneously given would provide insight. Two recent studies suggest that the GnRHa-related acute bone loss can be partially prevented by concomitant treatment with norethindrone (Riis et al. 1990; Survey and Judd 1992).

3.6 Progesterone and Progestins as Clinical Therapy

Despite the cumulative evidence that progesterone and progestins have potential positive effects on bone formation, few clinical data are available to support their therapeutic efficacy. One recent study (Gallagher et al. 1991) randomized women who were 3–6 years after menopause with medroxyprogesterone or placebo 10 mg on days 1–23 out of a 28-day cycle and measured bone changes density over 2 years. The data suggest that medroxyprogesterone treatment decreases the rate of bone loss in cortical but not in trabecular bone (Gallagher et al. 1991). No measurements of bone turnover were available.

The following two sections will present pilot data from our centre which indicate a role of progestins in bone formation. In these studies, bone density measurements have been systematically made while women were treated with cyclic medroxyprogesterone for hypothalamic amenorrhoea and when postmenopausal women with low bone densities and contraindications to oestrogen therapy were treated with continuous medroxyprogesterone.

3.6.1 Cyclic Progestin Treatment for Hypothalamic Amenorrhoea

Menstrual cycle disturbances are common in the population whether they are characterized by profound suppression of both oestrogen and progesterone production (e.g. amenorrhoea) or by normal oestrogen levels with minimal decreases in progesterone secretion (e.g. short luteal phases in cycles of normal interval). No treatment for hypothalamic amenorrhoea is known to increase trabecular bone density, although amenorrhoeic women who are anorectic or compulsive athletes who gain weight and recover their menstrual flow do have a partial regain in bone density (Bachrach et al. 1991; Lindberg et al. 1987). Because observational studies indicate that normal ovulatory function as well as normal menstrual cycle intervals are necessary to preserve peak bone mass during the premenopausal years (Prior et al. 1990), we postulated that replacement of cyclic progesterone might help bone loss. This section will review some early data suggesting that medroxyprogesterone given in a cyclic manner that equals the minimum normal luteal phase length (10 days) and in a dose similar to endogenous progesterone production (10 mg/day) has positive effects on bone balance in women with menstrual cycle disturbances.

Referred women whose menstrual cycles had stopped were evaluated to exclude pregnancy, ovarian failure, pituitary tumours and androgen excess. All were diagnosed as having hypothalamic amenorrhoea which had been present for more than 1 year. When menstrual flow remained absent in women with amenorrhoea for whom weight

Table 2. Demographic, clinical and quantitative computed tomography data in eight women with hypothalamic amenorrhoea prescribed cyclic medroxyprogesterone treatment

Subj. no.	Drug (mg/month)	Age (years)	BMI	Menstrual flow[a]	Exercise[b]	Initial QCT	Months between QCTs	Change over 12 months	Other problems[c]
1	100	21	18.7	13	1	119.7	12.0	+13.4	–
2	100	32	19.8	3	0	137.0	12.0	+11.3	–
3	100	47	26.0	4	0	111.0	7.5	+5.6	Diabetes type II diet
4	30	27	21.6	0	3	154.1	20.0	–5.6	–
5	50	31	16.5	14	0	107.1	19.0	–1.8	Addison's disease treatment
6	50	30	19.2	0	2	109.5	6.0	–7.6	–
7	0	21	17.6	0	2	98.2	12.0	–7.4	–
8	0	36	29.8	1	0	122.3	14.0	–7.6	Prolactin–slight ↑
Mean		31.8	21.2	4.4	0.9	119.9	12.8	0.045	
SD		9.1	4.5	5.8	1.2	18.0	4.9	8.8	

BMI, body mass index; QCT, quantitative computed tomography
[a] Episodes of menstrual flow during the year before the last QCT.
[b] Exercise categories: 0 = none, 1 = occas walking, 2 = 1/2 to 1 h of mild aerobics/week, 3 = > 1 h/week of aerobic exercise, 4 = > 2 h/week of intense training.
[c] Cortisone acetate 37.5 mg/day–physiological replacement; prolactin levels equal or are 1.5 times the top of the normal range.

had been normal or near normal for 1 year, each of the eight women was given a prescription for cyclic medroxyprogesterone with an explanation about its potential beneficial effects and minimal side effects (Kirkham et al. 1991). History of dietary and supplemental intake of calcium, adherence to recommended treatment, physical activity and menstrual flow were ascertained at 6-monthly visits.

Bone density measurements of T12 through L3 by QCT were performed initially when the prescription for cyclic medroxyprogesterone was given and were repeated a mean of 13 months (range 6–20) later. Using the previously described method, the normal QCT bone density in premenopausal women is 154 ± 22 mg/cm^3 ($n = 66$, $CV = 0.8\%$) (Prior et al. 1990). The eight enrolled women were a mean ($\pm$ SD) of 32 ± 9 years old (range 21–47 years) and had a mean body mass index (kg/m^2) of 21 ± 5 (Table 2). The initial QCT mean bone density in women with amenorrhoea was 120 ± 18 mg/cm^3 (range 98–154 mg/cm^3), which is significantly lower than normal (154 ± 22 mg/cm^3, $p = 0.0001$). Bone density measurements were adjusted to represent yearly change.

Some menstrual flow occurred in five women during the year preceding the second QCT measurement, and all except one of these women reported taking the prescribed or a partial dose of cyclic medroxyprogesterone (Table 2). No flow occurred in three women who reported taking ≤ 50 mg/month of medroxyprogesterone.

The three women who took the full dose of cyclic medroxyprogesterone had QCT spinal density increases of 13.4, 11.3 and 5.6 mg/cm^3 per year. Two women who did not take the medication had QCT losses of 7.4 and 7.5 mg/cm^3 per year. Partial doses of medication were taken by three women who had variable bone changes of –1.8, –5.6, and –7.6 mg/cm^3 per year. These data are shown as the annual percent change in bone density (Fig. 5).

By linear regression, the annual change in spinal bone density was significantly related to the dose of medroxyprogesterone ($R2 = .81$, $p = 0.002$). No other measured variable contributed significantly to the change in bone density. In a multiple regression model including both dose and flow, dose alone contributed 53% of the variance in bone change while flow contributed 2%.

These results indicate that women with secondary amenorrhoea who are of normal weight and have an adequate calcium intake can

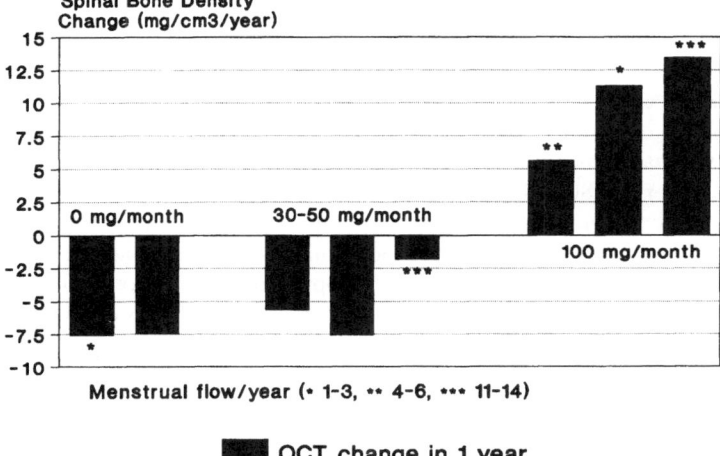

Fig. 5. Annual change in spinal bone density by quantitative computed tomography (QCT) in eight women with hypothalamic amenorrhoea who took varying monthly doses of medroxyprogesterone acetate although all were prescribed 10 mg/day to be taken for 10 days/month (e.g. 100 mg/month). The number of episodes of menstrual flow per year is indicated

have reversal of bone loss if treated with cyclic medroxyprogesterone. This pilot study needs to be repeated in a larger number of subjects in a controlled, blinded trial. It is likely that cyclic medroxyprogesterone treatment that produced normal luteal phase progesterone concentrations (or equivalent doses of norethindrone (Demarest et al. 1991) or promogestone (Tremollieres et al. 1992), but not levonorgesterol (Demarest et al. 1991) would have similar effects. Cyclic progestin treatment, however, may not be able to prevent bone loss during the first year or two after the onset of amenorrhoea. In this situation, activation of remodelling and the resorption phase dominate so that a larger dose and possibly longer duration of therapy may be necessary to prevent bone loss.

3.6.2 Cyclic Medroxyprogesterone for Athletic Women with Ovulatory Disturbances

It has been proposed that many women who experience fractures and develop osteoporosis after menopause had low bone densities before they became menopausal (Prior et al. 1991a). This postulate rests on the observed 80% 1-year prevalence of ovulation disturbances in a carefully selected and screened population of ovulatory urban women ages 20–42 years and on the association between anovulation and spinal bone loss (Prior et al. 1990). No population-based, prospective data are available to reliably indicate the normal variability of ovulation and luteal phase lengths; however, other studies also indicate a high prevalence of disturbances of ovulation (Vollman 1977; Nagata et al. 1986). Prospective studies in selected populations indicate that lack of ovulation can be clinically silent and commonly occurs in cycles that produce normal oestrogen levels (Soules et al. 1989; Prior et al. 1990).

On the basis of these observations, we set out to determine whether similar cyclic progestin treatment in women with less dramatic disturbances of menstrual cycle interval and ovulation would prevent bone loss. Because we were measuring bone by DXA methods, which include both cortical and trabecular spinal bone components, we chose to study exercising women. In this population we expected that weight bearing would be a consistent feature and therefore minimally influence the changes in DXA results. Preliminary results of a controlled, randomized, double-blind study of cyclic medroxyprogesterone treatment in 61 physically active premenopausal women, mean age 33 years (range 22–45) with abnormal menstrual cycles suggest that mean increases in spinal bone by DXA of 2%–3% per year can be expected (Prior et al. 1991b). The bone change was significantly related to cyclic medroxyprogesterone treatment ($p = 0.009$). These data suggest that, in athletic women with menstrual cycle disturbances ranging in severity from short luteal phase cycles to amenorrhoea, luteal phase replacement doses of medroxyprogesterone can promote bone formation.

3.6.3 Postmenopausal Women Treated
with Continuous Medroxyprogesterone

This pilot clinical study was designed to document whether, in a given postmenopausal woman, continuous medroxyprogesterone treatment would change the rate of bone loss. It also provides a feasibility study concerning the clinical use of medroxyprogesterone in menopausal women for whom oestrogen treatment is contraindicated or not desired. Six postmenopausal women from a clinical endocrine practice who were discovered to have low spinal trabecular bone densities were studied. All women for whom information on the same QCT instrument were available showing a rate of bone change when untreated and subsequently when on medroxyprogesterone therapy were included. Table 3 shows the clinical characteristics of these women and describes their other endocrine problems. The initial rate of bone change was measured over intervals of 6–15 months while they continued taking their usual hormonal medications and while they were supplemented to a total of 800 mg/day of calcium. They were then restudied during a 6–12 month period of treatment with medroxyprogesterone acetate (Provera R) 10 mg/day.

Trabecular bone density in the twelfth thoracic through the fourth lumbar spinal segments was measured using a single energy QCT with an edge-detection software program (Prior et al. 1990). This method has a coefficient of variation of 0.8% in premenopausal women but precision would likely approximate 1% in a postmenopausal population (Prior et al. 1990). Data were arithmetically adjusted to 12 months and reported as mg/cm^3 in K_2HPO_4 equivalents. Statistical analysis was by single sample (paired) Student's t-test with alpha conservatively preset at 0.05.

The six women ranging in age from 44 to 68 years were all postmenopausal, three surgically. Two women had treated breast cancer in remission, one woman had Hodgkin's disease and subsequent leukemia and two had hypopituitarism treated with thyroid and cortisone acetate in clinical doses and one with 1-thyroxine and cortisone acetate for primary thyroid and adrenal insufficiencies. One woman had stopped menstruating at the time of her pituitary surgery 32 years previously. One woman with histologically proven osteomalacia had been

Table 3. Open studies of spinal trabecular bone density changes (by QCT[a]) in postmenopausal patients with hormonal disturbances during untreated versus medroxyprogesterone acetate treatment periods

Subject	Baseline density (mg/cm^3)	Untreated period[b]		Treatment period[b]	
		Months apart	Bone change/year (mg/cm^3)	Months apart	Bone change/year (mg/cm^3)
1	26.7 ± 10	7	−9.8	6	3.6
2	108.9 ± 17	14	−7.9	7	−2.6
3	59.9 ± 11	6	−3.4	12	−2.8
4	60.4 ± 5	15	−3.7	8	−2.5
9	82.0 ± 5	6	−16.0	7	6.1
13	21.6 ± 6	8	−14.3	7	4.8
Mean	59.9	9.3	−9.2	7.8	1.1
SD	33.0	4.1	5.3	2.1	4.2

[a]QCT is quantitative computed tomography in mg/cm^3 K$_2$HPO$_4$, mineral equivalents.
[b]Change in bone density adjusted to a year, by paired t-testing; $p = 0.04$.

on a stable dose of cholecalciferol (50 000 IU/week) for 2 years before the initial rate of bone change was measured.

For each of the six women the initial bone density was significantly below the age-adjusted normal range – a mean of 60 mg/cm^3 while the normal mean for women averaging 56 years old would be approximately 110 mg/cm^3 (Table 3). Bone change over the 9.3 ± 4.1 months initially was unusually rapid, averaging a loss of 9.2 ± 5.3 mg/cm^3 per year ($p < 0.008$). However, during the 7.8 ± 2.1 months of the medroxyprogesterone treatment period, the bone change became positive though not significantly different from zero ($p = 0.55$). With each woman as her own control, the rate of bone change during medroxyprogesterone treatment was significantly different from that during the preprogestin phase ($p = 0.04$). These data are shown graphically in Fig. 6.

Although this is a very small open study, it suggests that medroxyprogesterone treatment was able to reverse the rapid bone loss in some women for whom standard treatment for postmenopausal osteopenia (oestrogen) was inappropriate. The women who continued to lose bone

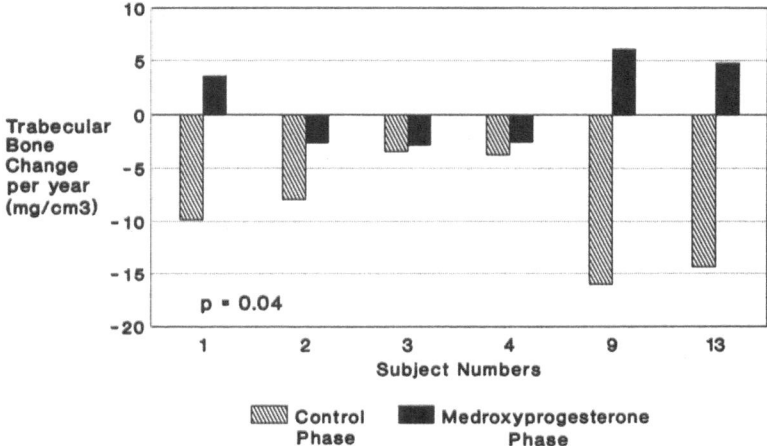

Fig. 6. The annual change in spinal bone density by quantitative computed tomography in six postmenopausal women with endocrine disease and contraindications for oestrogen treatment. The *cross-hatched bars* represent the change in bone adjusted to 1 year during treatment only with their stable and preexisting endocrine medications whereas the *black bars* represent the rate of change during 1 year of continuous medroxyprogesterone (10 mg/day) therapy. By paired statistics, the rate of change during the progestin treatment differed from the previous rate of change ($p = 0.04$)

were either taking glucocorticoids (patients 3 and 4) or were within the first months after premenopausal oophorectomy (patient 2). Because Grecu and colleagues (1991) have shown reversal of bone loss in steroid-treated asthmatic men on parenteral medroxyprogesterone, it is likely that a higher dose of medroxyprogesterone would be necessary to offset the glucocorticoid dose (which both patients 3 and 4 tended to double for minor symptoms).

Medroxyprogesterone may be an acceptable alternate therapy for low bone density in postmenopausal patients for whom oestrogen treatment is not feasible. Unlike other treatments that are effective for trabecular bone loss (such as calcitonin and bisphosphonates) medroxyprogesterone has an advantage in that it can also be used to treat menopausal symptoms such as vasomotor instability and vaginal atrophy (Lobo et al. 1984; Bullock et al. 1975; Schiff et al. 1980). The three patients with intact uteri did not have flow. The woman (patient 2) treated immediately after oophorectomy had excellent control of vasomotor symptoms. No patient complained of breast, bloating or mood symptoms.

Long term data on these six women would be desirable. However, two women (numbers 1 and 13) died of their malignancies in the year after the last bone density measurement. In addition, one patient (no. 2) has moved out of the region. Three women continue on medroxyprogesterone treatment with bone density measurements by DXA that are normal in one and just below the one standard deviation of age-matched normal values in two.

These pilot data on medroxyprogesterone treatment suggest it has the potential to prevent trabecular bone loss even in women with multiple medical problems. Controlled, blinded studies comparing medroxyprogesterone, oestrogen and oestrogen combined with cyclic progestin are feasible. Ideally, to also test the concept that the rate of bone turnover influences the net bone change during progestin treatment, the subjects should be characterized into two groups with either high or low bone remodelling rates. The women in each bone remodelling group could then be randomized to one of the three treatments. This would allow dissection of hormonal effects at varying extremes of bone turnover.

3.7 Progestin Combined with Oestrogens

There are at least three reasons to suspect that progestin given cyclically with oestrogen will enhance the effect on bone. The first is that oestrogen treatment leads to a decrease in both resorption and formation (Riggs et al. 1969) whereas oestrogen plus progestin treatment appears to stimulate bone formation (Christiansen et al. 1985). The second is that, although oestrogen withdrawal causes an increased rate of bone loss (Lindsay et al. 1978), sudden discontinuation of combined oestrogen and progestin therapy is associated with a normal rate of bone loss (Christiansen et al. 1981). Finally, we have previously shown that women with normal oestrogen productions needed cyclic endogenous progesterone secretion as well (i.e. a normal luteal phase length) to prevent spinal trabecular bone loss (Prior et al. 1990).

Although progestins have been advocated and used along with oestrogens for menopausal treatment for more than 20 years, the effects of the progestin in this combination have been assumed to be negligible. Even if treatment included progestins, that therapy was commonly called "oestrogen" treatment (Nachtigall et al. 1979). For these reasons, good human studies are scarce.

The majority of the evidence suggests that the two female gonadal steroids have synergistic and additive effects in bone metabolism. In cell culture, progesterone combined with oestrogen leads to a greater increase in DNA synthesis than either alone (Schevens et al. 1992). Careful studies in monkeys now indicate that the combined therapy is more likely than oestrogen alone to promote increased bone density or trabecular thickness (Jayo et al. 1990). Poorly controlled clinical studies also suggest that oestrogens and progestins have synergistic effects (McNeeley et al. 1991). Finally, low dose oestrogen with 10 mg/day of medroxyprogesterone appears to equal full dose oestrogen treatment, especially in cortical bone (Gallagher et al. 1991). In contrast, the 10-week-old rat treated with combined progesterone and supraphysiological doses of oestrogen shows loss of the beneficial effects of both oestrogen and progesterone when these two hormones are combined (Barbagallo et al. 1989). Given the above primate and human data, it is likely that the combined hormones act differently in the rat oophorectomy model. Further studies are needed.

3.8 Summary

This paper has briefly summarized the information relating to progesterone and bone remodelling that has been presented or published since 1990. The primary hypothesis posed by the new data is that progesterone treatment, in primates and humans, has its major effect on bone formation but little effect on resorption. Therefore, because of the time characteristics of coupled remodelling, when progesterone or progestin treatment is given during situations of high bone turnover, the net effect will be a continued loss of bone despite stimulated formation.

Several centres have now shown that progesterone added to osteoblast cultures in vitro causes receptor-mediated increases in DNA synthesis and skeletal alkaline phosphatase secretion as well as increased osteoblast numbers. There are recurrent and plausible connections between bone growth factors and progesterone's effects on the osteoblast. Growth factor interactions with progesterone will be an area within the bone field of increasing and fruitful study in the next few years. Studies now consistently show that osteocalcin and skeletal alkaline phosphatase are increased during treatment with progestins for osteoporosis. And finally, luteal phase increases in somatomedin C (also known as IGF-I), osteocalcin and skeletal alkaline phosphatase are consistent with the concept that bone formation is increased related to endogenous progesterone secretion in the normal menstrual cycle.

Oophorectomized women and animals experience marked early increases in bone remodelling. To understand the observed data, this review has postulated that the oophorectomized model is a good one in which to study gonadal steroid effects, provided that the rate of bone remodelling is characterized and the hormonal effects are seen in that context.

Finally, three different pilot clinical studies were briefly described. Data were presented showing that three progesterone-treated women who began therapy within a few days after premenopausal oophorectomy lost bone despite probably increased bone formation; however, two women of similar age and underlying diagnosis treated the same way but for whom surgery was many years previously experienced significant increases in bone density. The second pilot study indicated that women with hypothalamic amenorrhoea, when treated with cyclic medroxyprogesterone (10 mg/day for 10 days/month), experienced significant increases in QCT spinal bone density. The last study showed that continuous medroxyprogesterone treatment (10 mg/day) in women with multiple endocrine diseases for whom oestrogen treatment was not appropriate was associated with prevention of accelerated bone loss.

Continuing and accumulating evidence confirms a role for progesterone in bone formation. Because resorption is quicker and more efficient than formation within the bone remodelling cycle, if activation frequency is increased, despite evidence that progesterone treatment is increasing bone formation, the net change in bone density may be negative. Progesterone can continue to be understood as a bone-forming hormone but it must be seen to work within the context of the bone remodelling cycle.

Acknowledgements. These studies would not have been possible without the support and assistance of many people including Yvette Vigna, BA, RN, Arthur Burgess, PhD, John Wark MD, PhD, as well as those women who volunteered to participate in these studies.

References

Bachrach LK, Katzman DK, Litt IF, Guido D, Marcus R (1991) Recovery from osteopenia in adolescent girls with anorexia nervosa. J Clin Endocrinol Metab 72:602–606

Barbagallo M, Carbognani A, Palummeri E, Chiavaini M, Pedrazzoni M, Bracchi PG, Passeri M (1989) The comparative effect of ovarian hormone administration on bone mineral status in oophorectomized rats. Bone 10:113–116

Barengolts EI, Gajardo HF, Rusol TJ, D'Anza TJ, Pena M, Botsis J, Kukreja SE (1990) Effects of progesterone and post ovariectomy bone loss in aged rats. J Bone Miner Res 5:1143–1147

Biller BM, Coughlin JF, Saxe V, Schoenfeld D, Spratt DS, Klibanski A (1991) Osteopenia in women with hypothalamic amenorrhea: a prospective study. Obstet Gynecol 78:996–1001

Bullock JL, Massey FM, Gambrell RD jr (1975) Use of medroxyprogesterone acetate to prevent menopausal symptoms. Obstet Gynecol 46:165–168

Cann CE, Genant HK, Ettinger B, Gordan GS (1980) Spinal mineral loss in oophorectomized women. JAMA 244:2056–2059

Christiansen C, Christensen MS, Transbol (1981) Bone mass in post menopausal women after withdrawal of estrogen/gestagen replacement therapy. Lancet 1:459–461

Christiansen C, Nilas L, Riis BJ, Rodbro P, Deftos LJ (1985) Uncoupling of bone formation and resorption by combine oestrogen and progestogen therapy in postmenopausal osteoporosis. Lancet 2:800–801

Demarest KT, Jordan JJ, Hahn DW, Capetela RJ, Lau KH, Baylink D (1991) Direct stimulation of the proliferation and the differentiation of osteoblast-line cells by progestins. J Bone Miner Res 6:S139 (Abstract)

Demarest KT, Jordan JJ, Gunnet JW (1992) Stimulation of osteoblast-line cell proliferation by norethindrone which is not blocked by a progestin androgen or estrogen antagonist. J Bone Miner Res 7:S52 (Abstract)

Felsenfeld AJ, Iida-Kluin A, Hahn TJ (1992) Interrelationship between parathyroid hormone and insulin: effects on DNA synthesis in VMR-106–01 cells. J Bone Miner Res 7:1319–1325

Gallagher JC, Kable WT, Goldgar D (1991) The effect of progestin therapy on cortical and trabecular bone: comparison with estrogen. Am J Med 90:171–178

Grecu E, Simmons R, Baylink D, Haloran BP, Spencer ME (1991) Effects of medroxyprogesterone acetate on some parameters of calcium metabolism in patients with glucocorticoid-induced osteoporosis. Bone Miner 13:153–161

Horsman A, Simpson M, Kirby PA, Nordin BEC (1977) Non-linear bone loss in oophorectomized women. Br J Radiol 50:504–507

Jayo MJ, Weaver DS, Adams MR, Rankin SE (1990) Effects on bone of surgical menopause and estrogen therapy with or without progesterone replacement in cynomolgus monkeys. Am J Obstet Gynecol 614:618

Kalu DN, Salerno E, Liu CC, Echon R, Ray M, Garza-Zapata M, Hollis BW (1991) A comparative study of the actions of tamoxifen, estrogen and progesterone in the ovariectomized rat. Bone Miner 15:109–124

Karambolova KK, Snow GR, Anderson C (1986) Surface activity on the periosteal and corticoendosteal envelopes following continuous progestogen supplementation in spayed beagles. Calcif Tissue Int 38:239–243

Kirkham C, Hahn PM, Van Vugt DA, Carmichael JA, Reid RL (1991) A randomized, double-blind, placebo-controlled, cross-over trial to assess the side effects of medroxyprogesterone acetate in hormone replacement therapy. Obstet Gynecol 78:93–97

Lempert UG, Strong DD, Mohan S, Demarest K, Baylink DG (1992) Effect of progesterone on the mRNA levels of insulin-like-growth factors (IGFs), IGF-binding proteins (IGFBPs) and type-1 and type-2 IGFF receptors in human osteoblastic cells. In: Cohn DV, Gennari C, Tashjian AH Jr (eds) Calcium regulating hormones and bone metabolism. Elsevier, Amsterdam, pp 239–243

Lindberg JS, Powell MR, Hund MM, Ducey DE, Wade CE (1987) Increased vertebral bone mineral in response to reduced exercise in amenorrheic runners. Western J Med 146:39–42

Lindsay R, Hart DM, MacLean A (1978) Bone response to termination of oestrogen treatment. Lancet 1:1325–1327

Lobo RA, McCormick W, Singer F, Roy S (1984) Depo-medroxyprogesterone acetate compared with conjugated estrogens for the treatment of postmenopausal women. Obstet Gynecol 63:1–5

LoCascio V, Bonucci E, Imbimbo B, Ballanti P, Adami S, Milani S, Tartarotti D, Della Rocca C (1990) Bone loss in response to longterm glucocorticoid therapy. Bone Miner 8:39–51

Manolagas SC, Anderson DC (1978) Detection of high affinity glucocorticoid binding in the rat bone. J Endocrinol 76:379–380

Masuda A, Guo JZ (1991) Progesterone stimulates the secretion of a 35 kDa IGF-I binding protein from human osteoblast-like cells. Endocrine Society Abstract A1059

Mazzuoli GF, Minisola S, Bianchi G, Pacitti T, Russo R, Romagnoll E, Carnevale V, Scarneulina L, Celi FS, Ortore V, Carenza L (1990) The effects of oophorectomy on skeletal metabolism. Steroid Biochem Mol Biol 37:457–459

McNeeley SG, Schinfeld JS, Stovall TG, Ling FW, Buxton BH (1991) Prevention of osteoporosis by medroxyprogesterone acetate in post-menopausal women. J Gynaecol Obstet 34:253–256

Meema HE, Meema S (1968) Prevention of postmenopausal osteoporosis by hormone treatment of menopause. Can Med Assoc J 99:248–251

Mohan S, Baylink D (1991) Bone growth factors. Clin Orthop Related Res 263:30–48

Nachtigall LE, Nachtigall RH, Nachtigall RD, Beckmann EM (1979) Estrogen replacement therapy 1: a 10-year prospective study in the relationship to osteoporosis. Obstet Gynecol 53:277–281

Nagata 1, Kato K, Seki K, Furuya K (1986) Ovulatory disturbances. Causative factors among Japanese student nurses in a dormitory. J Adolescent Health Care 7:1–5

Nielsen HK, Brixen K, Bouillon R, Mosekilde L (1990) Changes in biochemical markers of osteoblastic activity during the menstrual cycle. J Clin Endocrinol Metab 70:1431–1437

Ohta H, Maketa K, Suda Y, lkeda T, Masuzawa T, Nozawa S (1992) Influence of oophorectomy on serum levels of sex steroids and bone metabolism and assessment of bone mineral density in lumbar trabecular bone by QCT-C value. J Bone Miner Res 6:659–665

Parfitt AM (1984) The cellular basis of bone remodelling: the quantum concept re-examined in light of recent advances in the cell biology of bone. Calcif Tissue Int 36:S37-S45

Prior JC (1990) Progesterone as a bone-trophic hormone. Endocr Rev 11:386–398

Prior JC, Vigna YM, Schechter MT, Burgess AE (1990) Spinal bone loss and ovulatory disturbances. N Engl J Med 323:1221–1227

Prior JC, Vigna YM, Alojado N (1991a) Progesterone and the prevention of osteoporosis. Can J Ob/Gyn & Women's Health Care 3:178–184

Prior JC, Vigna YM, Lentle BC, Rexworthy C, Connell D (1991b) Cyclic progestin and calcium increase spinal bone density in women athletes with menstrual cycle disturbances. Endocrine Society Abstract 450

Reed MJ, Christodoulides A, Kaistman R, Seppala M, Teale JD, Ghilchik MW (1992) The effect of endocrine therapy with medroxyprogesterone acetate, 4-hydroxy androstenedione and tamoxifen on plasma concentrations of insulin-like growth factor (IGF)-1, IGF-II and IGF BP-II in women with advanced breast cancer. Int J Cancer 52:208–212

Reeve J (1987) Bone turnover and trabecular plate survival after artificial menopause. Br Med J 295:757–760

Riggs BL, Jowsey J, Kelly PJ, Jones JD, Maher FT (1969) Effect of sex hormones on bone in primary osteoporosis. J Clin Investigation 48:1065–1072

Riis BJ, Christiansen C, Johansen JS (1990) Is it possible to prevent bone loss in young women treated with LHRH agonists? J Clin Endocrinol 70:920–924

Schevens BAA, Damen CA, Verhaar HJJ, Duursma SA (1992) Estrogen and progestins stimulate growth of normal human osteoblast-like cells in vitro. J Bone Miner Res 7 [Suppl 1]:PS S224

Schiff I, Tulchinsky D, Cramer D, Ryan KJ (1980) Oral medroxyprogesterone in the treatment of postmenopausal symptoms. JAMA 244:1443–1445

Smith ML, Fogelman I, Hart DM, Scott E, Leggate I (1989) Effect of etidronate disodium on bone turnover following surgical menopause. Calcif Tissue Int 44:74–79

Soules MR, McLachlan RI, Marit EK, Dahl KD, Cohen NL, Bremner WJ (1989) Luteal phase deficiency: characterization of reproductive hormones over the menstrual cycle. J Clin Endocrinol Metab 69:804–812

Stepan JJ, Pospichal J, Presl J, Pacovsky V (1987) Bone loss and biochemical indices of bone remodelling in surgically induced postmenopausal women. Bone 8:279–284

Stepan JJ, Pospichal J, Vratislav S, Kanka J, Mensik J, Presl J, Pacovsky V (1989) The application of plasma tartrate-resistant acid phosphatase to assess changes in bone resorption in respone to artificial menopause and its treatment with estrogen or norethisterone. Calcif Tissue Int 45:273–280

Survey ES, Judd HL (1992) Reduction of vasomotor symptoms and bone mineral density loss with combined norethindrone and long-acting gonadotropin-releasing hormone agonist therapy of symptomatic endometriosis: a prospective randomized trial. J Clin Endocrinol Metab 75:558–563

Tertinegg L, Heersche JN (1992) Progesterone stimulates bone nodule formation in rat calvarial cell cultures while estrogen has no effect. J Bone Miner Res 7 [Suppl 1]:S220

Tremollieres FA, Mohan S, Strong DD, Pouilles JM, Varin C, Baylink D, Ribot C (1991) Promogestone (PM) prevents bone loss in postmenopausal patients in a double blind clinical trial and acts as a mitogen in normal human bone cells (HBC) in vitro. J Bone Miner Res 6:S301(Abstract)

Tremollieres FA, Strong DD, Baylink D, Mohan S (1992) Progesterone and promogestone stimulate human bone cell proliferation and insulin-like growth factor 2 production. Acta Endocrinol 126:329–337

Vollman RF (1977) The menstrual cycle. In: Friedman EA (ed) Major problems in obstetrics and gynecology, vol 7. Saunders, Toronto

4 The Effect of Sex Steroids on Calciotropic Hormones In Vivo

Justin Silver, Ayal Epstein, Gideon Almogi,
and Tally Naveh-Many

4.1 Introduction

The expression of the parathyroid hormone (PTH) and calcitonin genes are both markedly decreased by 1,25-hydroxyvitamin D_3 ($1,25(OH)_2D_3$) both in vivo in rats (Silver et al. 1986; Naveh-Many and Silver 1988, 1990) and in vitro (Silver et al. 1985; Okazaki et al. 1988; Cote et al. 1987). The PTH gene's expression is also regulated by calcium, where the major effect is that of a low calcium increasing PTH mRNA levels in vivo (Rodriguez et al. 1991; Naveh-Many and Silver 1990). In contrast, calcitonin gene expression is not regulated by calcium in vivo (Rodriguez et al. 1991). At the level of secretion, calcium is the major regulator for both PTH, inversely, and calcitonin, directly.

The major metabolic bone disease is osteoporosis, and in post-menopausal osteoporosis estrogen replacement therapy is the most effective treatment. However, the mechanism of estrogen effect on bone remains to be fully explained. Part of the effect is in all probability due to a direct effect of estrogens on bone. Osteoblasts have been shown to have estrogen receptors (ER; Eriksen et al. 1988; Komm et al. 1988), and the addition of estrogens to bone cell lines has been shown to have functional effects (Fukayama and Tashjian 1989; Gray et al. 1987; Ernst et al. 1988), including the synthesis of growth factors such as transforming growth factor β (TGFβ), and insulin-like growth factors I and II (IGF-I and -II) (Ernst et al. 1989; Mohan et al. 1988; Rodan 1991). These findings show that estrogens act directly on bone. Estrogens might also have an indirect effect on bone by regulating the production of the calcium-regulating hormones. In vitro studies have shown that estrogens increase the secretion of PTH from both bovine and human parathyroid tissue and increase the secretion of calcitonin from a medullary carcinoma cell line (Greenberg et al. 1986, 1987; Duarte et al. 1988; Backdahl et al. 1991). We have recently studied the regulation of PTH and calcitonin gene expression in vivo in the rat and the presence of the ER mRNA and protein in the rat parathyroid and C cells (Naveh-Many et al. 1992).

4.2 Methods

4.2.1 Animals

Female rats of the Hebrew University strain weighing 150–170 g were anesthetized by pentobarbitol and bilateral ovariectomies performed. Other rats had sham operations performed. The rats were maintained on a normal diet for 2 weeks. The rats were then divided into groups of four and were given either no treatment (sham and ovariectomized) or 17β-estradiol (37–145 nmol/day). The β-estradiol (βE_2) was given either as a single injection i.p., as daily injections i.p. for 7 days, or as minipumps for 1 or 2 weeks in a much lower dose of 12 pmol/day. The rats were then anesthetized, blood samples taken, and the thyroparathyroid tissue excised under pentobarbital anesthesia, snap frozen in liquid nitrogen, and stored at –70°C until extraction. The rat uteruses were removed and weighed.

4.2.2 Measurement of Cellular mRNA Levels

RNA was extracted from rat thyroparathyroid tissue or individual wells of bovine parathyroid cells and the levels of PTH mRNA were measured by northern blots after extraction with RNAzol (Biotex, Houston, Texas). RNA was denatured and ethidium bromide was added to each sample at a concentration of 0.1 mg/ml. The samples were size-fractionated by electrophoresis on 1.25% agarose gels containing formaldehyde and transferred to Hybond filters (Amersham, UK) by diffusion blotting. The integrity of the RNA and the uniformity of RNA transfer to the membrane were determined by ultraviolet (UV) visualization of the ribosomal RNA bands of the gels and the filters. The filters were fixed by UV crosslinking and hybridized as previously described (Silver et al. 1986; Naveh-Many et al. 1992).

4.2.3 Polymerase Chain Reaction

Polymerase chain reaction (PCR) for the ER gene in rat thyroparathyroid tissue RNA extracts was performed (Saiki et al. 1988; Kawasaki 1990). RNA was reverse-transcribed into first strand cDNA using a kit (Amersham). PCR amplification of the cDNA was performed as described (Naveh-Many et al. 1992), 100–200 ng each of the upstream and downstream oligonucleotides specific to the rat ER 3' end of the gene (Koike et al. 1987). The size of the radiolabeled amplified DNA fragment was consistent with the distance between the primers.

4.2.4 Immunohistochemistry

Immunohistochemistry for the ER was performed on formalin-fixed paraffin-embedded tissues by the method of O'Keane et al. (O'Keane et al. 1990) as described (Naveh-Many et al. 1992). The primary antiserum was rabbit antiestradiol (Diagnostic Products, Los Angeles,

Calif.) or normal rabbit serum as a negative control; the bridging anti-body was biotinylated swine antiserum to rabbit immunoglobulin (DAKO, Denmark), and the slides were stained with peroxidase-labeled avidin (DAKO Quik Staining Kit 40). The slides were counter-stained with hematoxylin. Immunohistochemistry was performed for calcitonin with a polyclonal antibody (DAKO, Denmark). The immu-nocytochemical assay of ER with anti-17β-estradiol antibody with nu-clear staining only, and not cytoplasmic staining, has been shown to be an accurate and specific method for the determination of ERs (O'-Keane et al. 1990). The studies were repeated with 3-amino-9-ethyl-carbazol as the final stain instead of diaminobenzidine (DAB).

4.2.5 In Vitro Studies

Bovine parathyroid glands were washed in ethanol at the slaughter-house and transported to the laboratory in Hank's solution at 4°C. Cells were prepared by collagenase digestion of the glands. The cells were dispersed in M-199 medium with L-glutamine, 10 mM HEPES and 15% charcoal stripped newborn calf serum, counted in a hemocy-tometer, and plated in a 24-well tissue culture plate (Costar, Cam-bridge) at a density of 1×10^6 cells per well. The cells were cultured at 37°C in 5% CO_2 for 24 h to allow attachment and then the medium was changed to phenol red free defined medium (DCCM-1) (Biologi-cal Industries, Beth Haemek, Israel).

4.3 Results

Ovariectomy led to a small, nonsignificant decrease in PTH and calci-tonin mRNA levels (Figs. 1–3). PTH mRNA and calcitonin mRNA le-vels were markedly increased by 17β-estradiol (Figs. 1–3). There was no difference for three control genes studied, namely, actin mRNA (Fig. 1) which is derived from the total thyroparathyroid tissue, soma-tostatin mRNA, which is specific for the C cells in the thyroparathy-roid tissue (Naveh-Many and Silver 1988), and the 1,25(OH)$_2$D$_3$ re-ceptor (VDR) mRNA levels, which is present in the parathyroid and C cells, but not in the thyroid follicle cells (Naveh-Many et al. 1990).

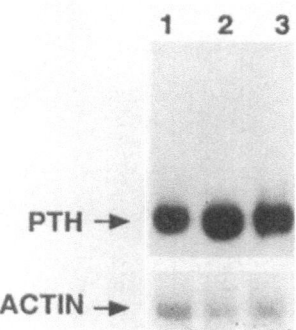

Fig. 1. Effect of estrogens on parathyroid hormone *(PTH)* mRNA in rat thyroparathyroid tissue. Agarose gel electrophoresis of RNA from a single rat hybridized for PTH mRNA and actin mRNA *(ACTIN)*. *Lanes: 1,* ovariectomized.; *2,* 17β-estradiol; *3,* sham. (From Naveh-Many et al. 1992)

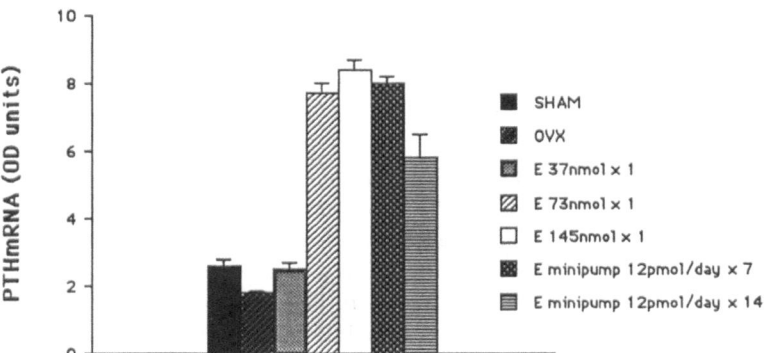

Fig. 2. Effect of estrogens on parathyroid hormone (PTH) mRNA in rat thyroparathyroid tissue. PTH mRNA levels derived from dot blots from sham-operated *(SHAM)*, ovariectomized *(OVX)*, and ovariectomized rats given 17β-estradiol *(E)*. Estradiol was given in single doses (× 1) of 37, 73, and 145 nmol, or 12 pmol/day as a continuous infusion by minipump for 7 or 14 days. *OD,* optical density. (From Naveh-Many et al. 1992)

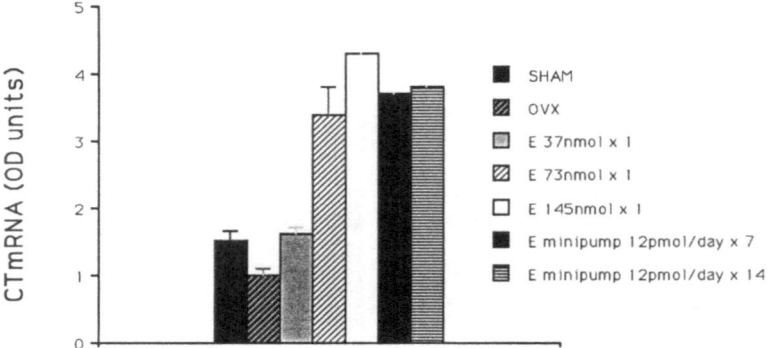

Fig. 3. Effect of estrogens on calcitonin mRNA in rat thyroparathyroid tissue. Calcitonin mRNA levels in sham-operated *(SHAM)*, ovariectomized *(OVX)* and estrogen-treated *(E)* rats. Results for dot blots are shown as the mean ± S.E.M. for four rats. *OD,* optical density. (From Naveh-Many et al. 1992)

The effect of estrogen on PTH mRNA levels was present 6 h after a 73-nmol dose of 17β-estradiol, but less marked than at 24 h. At 24 h single doses of 73 nmol and 145 nmol 17β-estradiol led to fourfold increases in mRNA levels for both PTH and calcitonin (Figs. 2, 3). Similar increases in mRNA levels were present after 17β-estradiol was given by osmotic minipump at a dose of 12 pmol/day for 7 or 14 days (Figs. 2, 3). There was no difference in serum calciums among the different groups of rats. 17β-Estradiol in these doses given to ovariectomized rats does not change serum 1,25-dihydroxyvitamin D_3 levels (Turner et al. 1987). These results demonstrate that estrogens regulate PTH and calcitonin gene expression in vivo. However, they do not demonstrate whether the estrogen directly affected the parathyroid and C cells, or indirectly, although one parameter of an indirect effect, serum calcium, did not change with ovariectomy or estrogen treatment.

Estrogen acts on its target organs by binding to a specific ER; therefore, in order to investigate whether rat thyroparathyroid tissue was a candidate target organ for estrogens we determined whether the ER mRNA was present (Koike et al. 1987). PCR with ER oligonucleo-

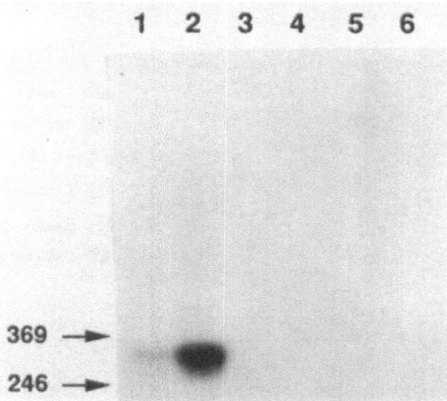

Fig. 4. Polymerase chain reaction products obtained by amplification of the rat estrogen receptor cDNA (ER cDNA) after reverse transcription of cellular RNA from female rat tissue were run on agorose gels and Southern blots hybridized with a ^{32}P-labeled estrogen receptor cDNA fragment. *Lanes: 1,* thyroparathyroid tissue; *2,* uterus; *3,* spleen; *4,* control without DNA; *5,* control without RNA; *6,* control without reverse transcriptase. These controls demonstrate that there was no contamination with nonspecific DNA, RNA or DNA in the RNA preparation, respectively. (From Naveh-Many et al. 1992)

tide primers using rat thyroparathyroid tissue RNA extracts showed a 300-bp band specific for the ER gene, indicating that it is expressed in this tissue (Fig. 4). It was also expressed in the rat liver (not shown), and rat uterus but not in rat spleen (Fig. 4).

When PCR was performed on Eco RI restricted rat DNA with the same oligonucleotide primers a band of 700 bp was demonstrated, which was larger than the 300-bp band that was predicted from the distance between the primers in the ERcDNA sequence. This larger PCR product presumably represents an intron and confirms that there was no DNA contamination of the PCR reaction.

The PCR product in the rat thyroparathyroid tissue might have been a product of the thyroid follicles, C cells and parathyroid cells, and it was therefore necessary to demonstrate that the ER was specific to the parathyroid and C cells, particularly because binding studies had not demonstrated an ER in parathyroid tissue (Prince et al. 1991). We did

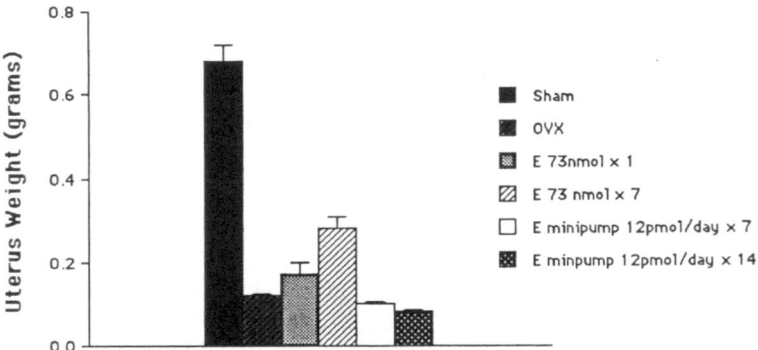

Fig. 5. Uterus weight in sham-operated *(SHAM)*, ovariectomized *(OVX)* and 17β-estradiol *(E)* treated rats; estradiol was given in a single dose (× 1), or daily for 7 days (× 7), or by minipump for 7 or 14 days (12 pmol/day). Results are shown as the mean ± SEM for four rats. (From Naveh-Many et al. 1992)

this by immunohistochemistry using a polyclonal antibody to 17β-estradiol, whose nuclear staining specifically labels the ER (O'Keane et al. 1990). As a negative control, normal rabbit serum was used instead of the primary antibody. As a positive control, rat uterus and the human ovary were used, where there was positive nuclear staining for the nuclear receptor. In the case of the human ovary, as expected, there was specific staining only of the granulosa cells. In the rat thyroparathyroid tissue the ER was found in the rat parathyroid and C cells, but not in the thyroid follicles, or in spleen (Naveh-Many et al. 1992). ER was also present in bovine parathyroid tissue. We also repeated these studies on both bovine parathyroid and rat thyroparathyroid tissue using 3-amino-9-ethyl-carbazol as the final stain instead of DAB with identical results. Together these results show that ER gene and its protein are expressed in the parathyroid and C cells.

   Another parameter of estrogen activity was measured in these rats, namely, uterus weight. The uterotrophic effect of estrogens is a well-characterized biological response to administered estrogens. As expected, ovariectomy leads to a large decrease in uterus weight, from a mean of 0.7 g to 0.15 g (Fig. 5). A pharmocological dose of 17β-estradiol (73 nmol) given daily for 7 days increased uterus weight twofold. A much smaller dose (12 pmol/day) given by a constant infusion pump for 2 weeks did not affect uterus weight (Fig. 5) despite its potent effect on PTH and calcitonin mRNA levels (Figs. 1–3). This small, physiologically relevant dose successfully separated an estrogen effect on the calcium regulating hormones from that on the uterus.

Preliminary studies show that testosterone given to castrated rats leads to an increase in PTH mRNA levels, but not as large an increase as that induced by estrogens. In addition preliminary in vitro studies show that both estrogens and progesterone added to bovine parathyroid cells lead to increased PTH mRNA levels.

4.4 Discussion

The action of estradiol to increase the expression of the PTH and calcitonin genes, and the presence of the ER mRNA and protein in the parathyroid and C cells, together with the earlier reports of a direct effect of estrogens to increase PTH and calcitonin secretion (Greenberg et al. 1986, 1987; Duarte et al. 1988; Backdahl et al. 1991) establish that these two organs are target organs for estrogens. This effect of estrogens might be important to the normal physiology of calcium homeostasis, by the action of PTH and calcitonin on their target organs, especially bone. In physiologic concentrations, PTH is anabolic to bone by its effect on osteoblasts, which combined with the effect of calcitonin to decrease osteoclastic bone resorption, would result in stronger bones.

The effect of estrogens in patients with osteoporosis was to increase serum PTH levels (Gallagher et al. 1980; Silverberg et al. 1989; Prince et al. 1990), which was postulated to be secondary to an estrogen-induced hypocalcemia (Gallagher et al. 1980; Prince et al. 1990), while the effect on calcitonin levels was variable (Reginster et al. 1989; Hurley et al. 1989), possibly due to the insensitivity of calcitonin immunoassays (Dick and Prince 1991). Our results, together with the earlier in vitro reports of a direct effect of estrogens on PTH and calcitonin secretion (Greenberg et al. 1986, 1987; Duarte et al. 1988; Backdahl et al. 1991), indicate that the effect of estrogens to increase PTH and calcitonin levels are a direct effect of estrogens on the parathyroid and C cells, respectively. In addition we have preliminary data with bovine parathyroid cells in primary culture, which show that estrogens in vitro increase PTH mRNA levels.

Ovariectomy produces osteopenia in female rats, which can be prevented by estrogen therapy (Takano-Yamamoto and Rodan 1990; Kalu et al. 1991) or by the administration of PTH (Liu and Kalu 1990) or diphosphonates (Wronski et al. 1991), which like calcitonin, inhibit bone resorption. In studies in osteoporotic patients PTH and calcitonin have been shown to have an anabolic effect on trabecular bone with an increase in vertebral bone density. These studies used PTH alone (Reeve et al. 1980) or PTH with $1,25(OH)_2 D_3$ (Slovik et al. 1986), or pulsatile PTH and sequential calcitonin (Hesch et al. 1989). Therefore, estrogens act on bone to prevent osteoporosis not only by a direct action on osteoblasts but also indirectly by its action on the parathyroid gland and the C cells. Osteoblasts have ERs (Eriksen et al. 1988; Komm et al. 1988), which are probably central to the effects of estrogen on bone, but it is intriguing that the body might use other mechanisms, such as an effect on PTH and calcitonin, to ensure normal bone strength.

17β-Estradiol given at low doses by minipump for 2 weeks increased PTH and calcitonin mRNA levels with no uterotrophic effect as measured by uterus weight, which suggests that it might be possible to develop dose regimes for estrogens or estrogen analogues with an effect on the calcium-regulating hormones but not on the female phenotypic organs. However, in the present study uterus weight was measured, and not the expression of uterus-specific genes, which might be a more relevant parameter. The clearer understanding of estrogen action will allow the development of treatment strategies of relevance to every postmenopausal woman. This is particularly important, because although postmenopausal estrogen replacement leads to a reduction of about 60% in hip fractures (Kiel et al. 1987), it may lead to a 1.3-fold increase in breast cancer (Colditz et al. 1990) and a sixfold increase in uterine cancer (Goldman and Tosteson 1991). There is, therefore, an urgent need for new organ specific estrogen compounds or dose regimes. The biological model reported here might allow the successful separation of estrogen's calciotrophic effect from its uterotrophic effect, which would be useful for the testing of such estrogen analogues.

Acknowledgements. This work was supported by grants from the National Institutes of Health (grant DK 38696), the United States–Israel Binational Science Foundation (BSF), and the German Israel Foundation (GIF).

References

Backdahl M, Howe JR, Lairmore TC, Wells SAJ (1991) The molecular biology of parathyroid disease. World J Surg 15:756–762

Colditz GA, Stampfer MJ, Willett WC, Hennekens CH, Rosner B, Speizer FE (1990) Prospective study of estrogen replacement therapy and risk of breast cancer in postmenopausal women. JAMA 264:2648–2653

Cote GJ, Rogers DG, Huang ES, Gagel RF (1987) The effect of 1,25-dihydroxyvitamin D3 treatment on calcitonin and calcitonin gene-related peptide mRNA levels in cultured human thyroid C-cells. Biochem Biophys Res Commun 149:239–243

Dick IM, Prince RL (1991) Transdermal estrogen replacement does not increase calcitonin secretory reserve in postmenopausal women. Acta Endocrinol (Copenh) 125:241–245

Duarte B, Hargis GK, Kukreja SC (1988) Effects of estradiol and progesterone on parathyroid hormone secretion from human parathyroid tissue. J Clin Endocrinol Metab 66:584–587

Eriksen EF, Clovard DS, Berg NJ, Graham ML, Mann KG, Spelsberg TC, Riggs BL (1988) Evidence of estrogen receptors in normal human osteoblast-like cells. Science 241:84–86

Ernst M, Schmid CH, Froesch ER (1988) Enhanced osteoblast proliferation and collagen gene expression by estradiol. Proc Natl Acad Sci USA 85:2307–2310

Ernst M, Heath JK, Rodan GA (1989) Estradiol effects on proliferation, messenger ribonucleic acid for collagen and insulin-like growth factor-I, and parathyroid hormone-stimulated adenylate cyclase activity in osteoblastic cells from calvariae and long bones. Endocrinology 125:825–833

Fukayama S, Tashjian AHJ (1989) Direct modulation by estradiol of the response of human bone cells (SaOS-2) to human parathyroid hormone (PTH) and PTH-related protein. Endocrinology 124:397–401

Gallagher JC, Riggs BL, Deluca HF (1980) Effect of estrogen on calcium absorption and serum vitamin D metabolites in postmenopausal osteoporosis. J Clin Endocrinol Metab 51:1359–1364

Goldman L, Tosteson ANA (1991) Uncertainty about postmenopausal estrogen. Time for action not debate. N Engl J Med 325:800–802

Gray TK, Flynn TC, Gray M, Nabell LM (1987) 17β-Estradiol acts directly on the clonal osteoblastic cell line UMR106. Proc Natl Acad Sci USA 84:6267–6271

Greenberg C, Kukreja SC, Bowser EN, Hargis GK, Henderson WJ, Williams GA (1986) Effects of estradiol and progesterone on calcitonin secretion. Endocrinology 118:2594–2598

Greenberg C, Kukreja SC, Bowser EN, Hargis GK, Henderson WJ, Williams GA (1987) Parathyroid hormone secretion: effect of estradiol and progesterone. Metabolism 36:151–154

Hesch RD, Busch U, Prokop M, Delling G, Rittinghaus EF (1989) Increase of vertebral bone density by combination therapy with pulsatile 1–38 hPTH and sequential addition of calcitonin nasal spray in osteoporotic patients. Calcif Tissue Int 44:176–180

Hurley DL, Tiegs RD, Barta J, Laakso K, Heath H (1989) Effects of oral contraceptive and estrogen administration on plasma calcitonin in pre- and postmenopausal women. J Bone Miner Res 4:89–95

Kalu DN, Liu CC, Salerno E, Hollis B, Echon R, Ray M (1991) Skeletal responses of ovariectomized rats to low and high doses of 17β-estradiol. Bone Miner 14:175–187

Kawasaki ES (1990) Amplification of RNA. In: Innis MA, Gelfand DH, Sninsky JJ, White TJ (eds) PCR protocols: a guide to methods and applications. Academic Press, New York, pp 21–27

Kiel DP, Felson DT, Anderson JJ, Wilson PWF, Moskowitz MA (1987) Hip fractures and the use of estrogens in postmenopausal women. N Engl J Med 817:1169–1174

Koike S, Sakai M, Muramatsu M (1987) Molecular cloning and characterization of rat estrogen receptor cDNA. Nucl Acids Res 15:2499–2513

Komm BS, Terpening CM, Benz DJ, Graeme KA, Gallegos A, Korc M, Greene GL, O'Malley BW, Haussler MR (1988) Estrogen binding, receptor mRNA, and biological response in osteoblast-like osteosarcoma cells. Science 241:81–84

Liu CC, Kalu DN (1990) Human parathyroid hormone (1–34) prevents bone loss and augments bone formation in sexually mature ovariectomized rats. J Bone Miner Res 5:973–982

Mohan S, Jennings JC, Linhart TA, Baylink D (1988) Primary structure of human skeletal growth factor: sequence homology with insulin-like growth factor-II. Biochim Biophys Acta:996:44–55

Naveh-Many T, Silver J (1988) Regulation of calcitonin gene transcription by vitamin D metabolites in vivo in the rat. J Clin Invest 81:270–273

Naveh-Many T, Silver J (1990) Regulation of parathyroid hormone gene expression by hypocalcemia, hypercalcemia, and vitamin D in the rat. J Clin Invest 86:1313–1319

Naveh-Many T, Marx R, Keshet E, Pike JW, Silver J (1990) Regulation of 1,25-dihydroxyvitamin D3 receptor gene expression by 1,25-dihydroxyvitamin D3 in the parathyroid in vivo. J Clin Invest 86:1968–1975

Naveh-Many T, Almogi G, Livni N, Silver J (1992) Estrogen receptors and biologic response in rat parathyroid tissue and C-cells. J Clin Invest 90:2434–2438

Okazaki T, Igarashi T, Kronenberg HM (1988) 5'-Flanking region of the parathyroid hormone gene mediates negative regulation by 1,25-(OH)2 vitamin D3. J Biol Chem 263:2203–2208

O'Keane JC, Okon E, Moroz K, Burke B, Sheahan K, O'Brien MJ (1990) Anti-estradiol immunoperoxidase labeling of nuclei, not cytoplasm, in paraffin sections, determines estrogen receptor status of breast cancer. Am J Surg Pathol 14:121–127

Prince RL, Schiff I, Neer RM (1990) Effects of transdermal estrogen replacement on parathyroid hormone secretion. J Clin Endocrinol Metab 71:1284–1287

Prince RL, MacLaughlin DT, Gaz RD, Neer RM (1991) Lack of evidence for estrogen receptors in human and bovine parathyroid tissue. J Clin Endocrinol Metab 72:1226–1228

Reeve J, Meunier PJ, Parsons JA, Bernat M, Bilvoet OLM, Courpron P, Edouard C, Klenerman L, Neer RM, Renier JC, Slovik D, Vismans FJFE, Potts JT (1980) Anabolic effects of human parathyroid hormone fragment on trabecular bone in involutional osteoporosis: a multicentre trial. Br Med J 280:1340–1344

Reginster JY, Deroisy R, Albert A, Denis D, Lecart MP, Collette J, Franchimont P (1989) Relationship between whole plasma calcitonin levels, calcitonin secretory capacity, and plasma levels of estrone in healthy women and postmenopausal osteoporotics. J Clin Invest 83:1073–1077

Rodan GA (1991) Mechanical loading, estrogen deficiency, and the coupling of bone formation to bone resorption. J Bone Miner Res 6:527–530

Rodriguez M, Felsenfeld AJ, Llach F (1991) Calcemic response to parathyroid hormone in renal failure: role of calcitriol and the effect of parathyroidectomy. Kidney Int 40:1063–1068

Saiki RK, Gelfand DH, Stoffel S, Scharf SJ, Iguchi R, Horn GT, Mullis B, Erlich HA (1988) Primer-directed enzymatic amplification of DNA with a thermostable DNA polymerase. Science 239:487–491

Silver J, Russell J, Sherwood LM (1985) Regulation by vitamin D metabolites of messenger ribonucleic acid for preproparathyroid hormone in isolated bovine parathyroid cells. Proc Natl Acad Sci USA 82:4270–4273

Silver J, Naveh-Many T, Mayer H, Schmelzer HJ, Popovtzer MM (1986) Regulation by vitamin D metabolites of parathyroid hormone gene transcription in vivo in the rat. J Clin Invest 78:1296–1301

Silverberg SJ, Shane E, De La Cruz L, Segre GV, Clemens TL, Bilezikian JP (1989) Abnormalities in parathyroid hormone secretion and 1,25-dihydroxyvitamin D_3 formation in women with osteoporosis. N Engl J Med 320:277–281

Slovik DM, Rosenthal DI, Doppelt SH, Potts JT, Daly MA, Campbell JA, Neer RM (1986) Restoration of spinal bone in osteoporotic men by treatment with human parathyroid hormone (1–34) and 1,25-dihydroxyvitamin D. J Bone Miner Res 1:377–381

Takano-Yamamoto T, Rodan GA (1990) Direct effects of 17β-estradiol on trabecular bone in ovariectomized rats. Proc Natl Acad Sci USA 87:2172–2176

Turner RT, Vandersteenhoven JJ, Bell NH (1987) The effects of ovariectomy and 17β-estradiol on cortical bone histomorphometry in growing rats. J Bone Miner Metab 2:115–122

Wronski TJ, Yen CF, Scott KS (1991) Estrogen and diphosphonate treatment provide long-term protection against osteopenia in ovariectomized rats. J Bone Miner Metab 6:387–394

5 Effects of Estrogen on Growth Factors in Bone

Lynda F. Bonewald

5.1 Introduction

Post menopausal bone loss occurs when coupling between bone formation and bone resorption is no longer balanced. Estrogen replacement therapy results in decreased bone turnover, decreased bone resorption, and decreased fracture occurrence in postmenopausal osteoporosis. While in vivo experiments and clinical trials clearly show that estrogen treatment results in inhibition of bone resorption and new bone formation, in vitro experiments have not provided clear-cut clues to the mechanisms by which estrogen exerts its effects on bone. Data suggest that estrogen may act on bone by several mechanisms, either directly through estrogen receptors, indirectly through other cells besides bone cells, and in concert with other factors and hormones such as the progesterones.

5.2 Relationship of Estrogen to Bone Growth Factors

Direct delivery by infusion of 17β-estradiol into the femur of ovariectomized rats lead to increased bone formation and restored bone loss due to the ovariectomy (Takano-Yamamoto and Rodan 1990). In vitro experiments have shown that estrogen can stimulate responses in chondrocytes and osteoblasts that are assumed to be representative of the bone formation response such as an increase in alkaline phosphatase, collagen production, etc.

In chondrocytes, estrogen inhibits cell proliferation (Takahashi and Noumura 1987; Scranton et al. 1975), stimulates collagen production as assayed by S^{35} incorporation (Blanchard et al. 1991), and increases alkaline phosphatase (Carrasocsa et al. 1981).

Both estrogen and tamoxifen stimulated prostaglandin synthesis in rabbit articular chondrocytes (Rosner et al. 1983). In cartilage, a sex-dependent response to estrogen may exist as suggested in two recent studies, one in which estrogen showed a direct and sex-specific enhancement of endochondral bone formation in fetal mouse long bones (Schwartz et al. 1991) and a separate study in which estrogen increased collagen production only in chondrocytes derived from adult female rats (Nasatzky et al. 1993). These studies suggest that the effects of estrogen on chondrocytes are sex-specific and dependent on the differentiation stage of the chondrocytes.

Estrogen receptors have been demonstrated on osteoblast cells (Komm et al. 1988; Eriksen et al. 1988) and estrogen clearly increases α(I) procollagen mRNA in rat primary osteoblasts (Ernst et al. 1987, 1989a) and in osteoblast cell lines (Komm et al. 1988; Ernst et al. 1987, 1989b). Reports of the effects of estrogen on osteoblast cell proliferation have been contradictory and confusing. Estrogen stimulates proliferation in fetal rat calvarial cells and trabecular bone cells (Ernst et al. 1987, 1989a). The response was abolished by the antiestrogen tamoxifen and proliferation was mediated by transcription of insulin-like growth factor-I (IGF-I). In contrast, estrogen inhibited proliferation in an osteoblastic cell line UMR-106, (Gray et al. 1987a,b; Bankson et al.

1989) which was abolished by pretreating the cells with $1,25(OH)_2D_3$ which was assumed to induce differentiation (Gray et al. 1987a), and inhibited other osteosarcoma cell lines (Scranton et al. 1975). Estrogen was found to have no direct effect on proliferation and differentiation of normal human osteoblast-like cells (Keeting et al. 1991). A recent study utilizing a procedure to examine proliferation and expression of the osteoblast phenotype of single cells found that estrogen was not mitogenic for human osteoblasts (Rickard et al. 1993). Estrogen did not potentiate the actions of transforming growth factor-β (TGF-β) or interleukin-1 (IL-1) on proliferation by these cells. These differences could be due to culture conditions or differences in stages of osteoblast differentiation.

Estrogen enhances IGF-I mRNA in osteoblasts from rat calvaria and long bones (Ernst et al. 1989a). In UMR-106, estrogen stimulates the production of alkaline phosphatase and the bone growth factors, IGF-I, IGF-II and TGF-β (Gray et al. 1989b). Clearly the production of TGF-β is enhanced by estrogen in a number of osteoblast cell types such as ROS 17/2.8 (Komm et al. 1988) and in normal human explant bone cells (Ousler et al. 1991). Estrogen also enhances other parameters associated with bone formation such as vitamin D receptors in ROS 17/2.8 cells, which were abolished by tamoxifen treatment (Liel et al. 1992).

Alternatively, estrogen appears to inhibit factors and cells associated with bone resorption. Estrogen inhibited parathyroid hormone (PTH)-stimulated prostaglandin production and PTH-stimulated bone resorption (Pilbeam et al. 1989). Several studies have shown that estrogen decreases PTH-stimulated bone resorption in postmenopausal women (Gallagher and Wilkinson 1973; Selby and Peacock 1986). Estrogen appears to inhibit the production of cytokines with catabolic effects on bone such as IL-1, IL-6 and tumor necrosis factor (Girasole et al. 1992; Ralston et al. 1990; Pacifici et al. 1989; Stock et al. 1989). IL-1 and IL-6 are released by blood cells after oophorectomy (Pioli et al. 1992).

Estrogen inhibits osteoclast formation in vitro probably by decreasing IL-6 production by stromal cells (Jilka et al. 1992). Estrogen appears to have direct effects on the osteoclast and was shown to inhibit pit formation on bone slices (Ousler et al. 1993). As osteoclasts produce large amounts of IL-6, future studies may find that estrogen inhibits the production of this factor by osteoclasts as well as by osteoblasts.

Growth factors, in turn, may regulate the production of estrogen by bone cells. It has been suggested that tissue concentrations of sex steroid hormones may be more important in the postmenopausal female, especially the concentration of estrogens in bone tissue instead of plasma (Purolit et al. 1992). Bone cells may contain the enzymes necessary to convert sex steroid precursors to estrogen and Purolit and coworkers suggest that growth factors could regulate these enzymes in bone cells. These are based on observations by other investigators (Singh and Reed 1991; Reed et al. 1992) showing that IGF-I, IGF-II, IL-1 and IL-6 will regulate these estrogenic enzymes in breast cancer cells. Such studies remain to be performed using bone cells.

5.3 Effects of Estrogen in Combination with Progestins

The bone loss occurring during postmenopausal osteoporosis may not be completely due to a lack of estrogen but due to a lack of the combination of estrogen plus progesterone. Estrogen appears more effective in preventing bone loss which results in a reduction of bone formation but this leads to a constant bone mass (Christiansen et al. 1985). Progestogens may increase bone formation (Lindsay et al. 1978) and the combination of estrogen plus progestogen appears to result in increased bone mineral content by uncoupling resportion with formation (Christiansen et al. 1985). These observations have generally been found to be true in humans (Christiansen and Riis 1990) and rodents (Kerhl 1990; Barbagallo et al. 1989; Barengolts et al. 1990) Bain and coworkers (1993) found that high-dose gestagens enhanced estrogen induced bone formation in the ovariectomized mouse. They found that a 17α-hydroxyprogesterone, megestrol acetate, had no effect on vertebral bone resorption but increased tibial bone resorption. Treatment with a 19-nortestosterone, noresthisterone acetate, decreased bone resorption. Surprisingly, norethisterone acetate plus estrogen increased endosteal bone formation 36% over estrogen alone and megestrol acetate plus estrogen increased endosteal bone formation 2.3-fold. These studies show that gestagens may play an important role in regulating the actions of estrogens on bone tissue. These investigators concluded that gestagens have the capacity to enhance the actions of estrogen by augmenting osteoblast recruitment and activity. Alternatively, the synergistic interaction could be due to the conversion of gestagens to estrogenic forms, but this theory was dismissed as gestagens have nonestrogenic skeletal effects. The different effects of megastrol acetate and noresthisterone acetate in these studies support observations that different gestagens, at differing amounts, alone and in combination, can yield dramatically different responses. Generalizations concerning the actions of gestagens alone and in combination with estrogen should not be made without further study.

5.4 Effects of the Antiestrogen Tamoxifen on Bone

The antiestrogen tamoxifen, used to treat patients with breast cancer, appears to have effects similar to estrogen on bone. Though generally regarded as an estrogen antagonist, tamoxifen can have agonist activity. In experimental animal studies, tamoxifen inhibits bone turnover and inhibits loss of trabecular bone in oophorectomized rats (Jordan et al. 1987; Turner et al. 1987, 1988). In a number of studies examining the effects of tamoxifen on breast cancer it was noted that these patients retained bone mineral content (Ryan et al. 1991; Fentiman et al. 1989; Fornander et al. 1990; Love et al. 1988; Tarken et al. 1989). In one study using 140 postmenopausal women with breast cancer, tamoxifen was shown to stop the loss of bone from the lumbar spine but not from the radius (Love et al. 1992). Very little is known concerning the mechanism whereby tamoxifen prevents bone loss.

5.5 The Insulin-Like Growth Factors

IGF-I and -II are present in high levels in bone, with IGF-II considered to be the major IGF in bovine and human bone (Mohan et al. 1988; Linkhart et al. 1986). PTH stimulates release of IGF-I and -II from neonatal mouse calvaria (Linkhart and Mohan 1989) much like TGF-β is released from resorbing bone (Pfielschifter and Mundy 1987). As was postulated with TGF-β, it was also postulated that the IGFs might be released by osteoclastic resorption and also be coupling agents be-

tween resorption and formation (Farley et al. 1987). The IGFs are produced by osteoblast cells (Wergedal et al. 1986) and stimulate DNA
and protein synthesis in rat calvarial cells (Schmidt et al. 1983; Canalis
1980) and in human bone cells (Wergedal et al. 1990). In vitro estrogen stimulates secretion of IGF by UMR-106 cells (Gray et al.
1989b). In vivo administration of IGF-I increases bone density or prevents bone loss in ovariectomized rats (Spencer et al. 1991; Ammann
et al. 1991; Mueller and Cortesi 1991; Kalu et al. 1991). The effects of
growth hormone on bone growth may be direct or mediated through
IGF (Nilsson et al. 1990) and growth hormone has been shown to
potentiate the effects of IGF-I on chondrocyte colony formation (Lindahl et al. 1987).

The binding proteins for the IGFs modulate their biological activity
and these binding proteins may also be regulated by estrogen. IGF-
BP1 is a 30-kDa circulating protein regulated by insulin and can inhibit or potentiate IGF-I activity depending on the cell system
(McCusker et al. 1989; Clemmons 1990). IGF-BP2 is a 34-kDa protein
predominantly found in cerebrospinal fluid and has a greater affinity
for IGF-II than for IGF-I (Baxter and Martin 1989). Levels of this 34-
kDa BP were found to be significantly greater in elderly women than
in young women (Donahue et al. 1990). IGF-BP3, 53 kDa, is regulated
by growth hormones and IGF-I (Baxter and Martin 1986; Rosenfeld et
al. 1990). Purified IGF-BP3 can inhibit or stimulate fibroblast proliferation depending on culture conditions (De Mellow and Baxter 1988).
Estradiol has been shown to regulate IGF-BP3 production by fetal rat
calvaria cells (Schmid et al. 1989). Human bone cells produce both
IGF-BP3 and -4. IGF-BP4, 24 kDa, is produced by the bone cells,
MC3T3 cells (Scharla et al. 1991) and TE85 cells, and is distinctly different from the previous IGF-BPs but with significant amino and car-
boxy terminal sequences (La Tour et al. 1990). IGF-BP4 binds with
high affinity to both IGF-I and -II to inhibit their proliferation activity.
Complex relationships exist between IGFs and their binding proteins
(Froger-Gaillard et al. 1989) and these may be complicated with estrogen treatment.

5.6 The Bone Morphogenetic Proteins

The earliest demonstration of bone-inducing factors was the pioneering work of Marshall Urist (1965) in which explants from demineralized bone induced new bone formation when implanted into ectopic sites in rodents. These factors were difficult to purify due to low abundance in bone and the cumbersome nature of the osteoinductive bioassay. The clever approach which resulted in identification of BMP 2–7 was digestion of highly purified but not pure osteogenic material for peptides which were sequenced and cloned and the resulting recombinant material bioassayed (Wozney et al. 1988). A series of proteins were identified termed BMP 1–7, all of which possessed osteoinductive activity except BMP 1. BMP 1 is homologous to serine proteinase and may play a role in processing of the true BMPs (Rawl-

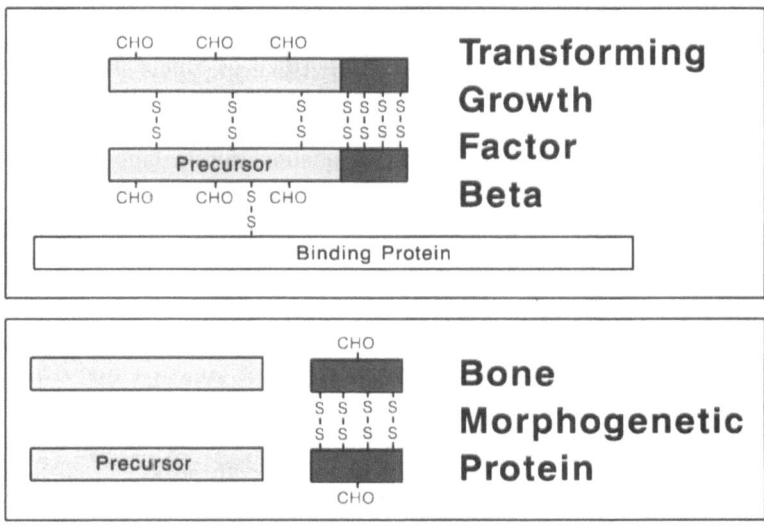

Fig. 1. Comparison of the structure of transforming growth factor-β (TGF-β) to bone morphogenetic protein (BMP). TGF-β is noncovalently associated with the homodimeric precursor or latency-associated peptide. One of the precursor regions may or may not be covalently associated with a binding protein. The precursor for the BMPs is monomeric and probably does not confer latency to the molecule. *CHO*, Chinese hamster ovary cells

Table 1. Characteristics of BMP 2 and TGF-β1

Characteristic	BMP 2	TGF-β1
Ectopic bone formation	Yes	No
New bone formation	?	Yes
Osteocalcin production	Promotes	Inhibits
Mineralization	Promotes	Inhibits
Amount in bone	1–2 ng/g	450 ng/g
Latent forms	?	Yes
Mature form	30-kDa homodimer glycosylated	25-kDa homodimer nonglycosylated
Precursor	Monomer	Latency-associated homodimer
Binding proteins	?	Yes
Induced by estrogen	?	Yes-latent

BMP, bone morphogenetic protein; TGF-β1, transforming growth factor-β1

ings and Barrett 1990). A previously purified osteoinductive protein called osteogenin is identical to BMP 3 (Luyten et al. 1989).

The BMPs belong to the TGF-β superfamily and more closely resemble members involved in differentiation during embryogenesis such as *Drosophila decapentaplegic* and Vg-1 (Ozkaynak et al. 1990). In fact the BMPs are probably mammalian homologues to these differentiation molecules. Like TGF-β, the mature BMPs contain seven cysteines but unlike TGF-β, the 30-kDa BMP homodimers are active, glycosylated, and their propeptides contain no cysteine residues (except for BMP 3) and therefore are secreted as monomers when expressed as recombinant material in CHO cells (Wozney 1992) (see Fig. 1). The nature of storage or "latent" forms of BMPs in the content of bone or bone cells is unknown. BMPs have the opposite effects of TGF-β on a number of osteoblast-like cells with respect to alkaline phosphatase induction, but like TGF-β stimulate DNA and collagen synthesis and protein accumulation (Chen et al. 1991). Also, functionally unlike TGF-β these molecules induce osteocalcin expression (Yamaguchi et al. 1991) (see Table 1). Little is known concerning the effects of estrogen on BMP production.

5.7 Transforming Growth Factor-β

Bone is the largest storage site for TGF-β1 in the body although the most concentrated source is platelets (for review see Bonewald and Mundy 1990). Bone appears to contain unique latent forms of TGF-β that are not detectable in platelets (see Fig. 2) (Bonewald et al. 1991; Jennings and Mohan 1990; Wakefield et al. 1988). This suggests regulation, storage, and targeting differences between latent complexes derived from bone compared to platelets and other cell types. In vivo, TGF-β stimulates the production of new bone (Marcelli et al. 1990; Noda and Camilliere 1989) and in vitro TGF-β generally stimulates osteoblast function and inhibits osteoclastic resorption (see reviews,

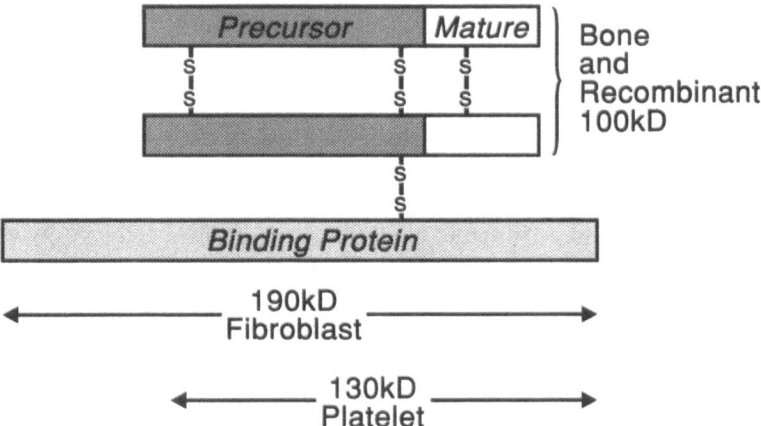

Fig. 2. Several latent transforming growth factor-β (TGF-β) complexes are produced by various cell types. Bone cells produce large amounts of the 100-kDa latent complex, which lacks the TGF-β binding protein (LTBP). Fibroblasts and some bone cells produce the 290-kDa complex which contains the LTBP. The latent complex in platelets contains a truncated 130-kDa form of the binding protein plus the 100-kDa portion. The binding protein for TGF-β is functionally and distinctly different from the binding proteins for the insulin-like growth factors (IGFs). The LTBP is produced within the cell and covalently attached to a precursor chain. LTBP does not confer latency to the complex. In contrast, IGF-BPs are secreted as distinct proteins which can then bind IGFs and neutralize or enhance their activity

Bonewald and Mundy 1990; Centrella et al. 1991). TGF-β will initiate chondrogenesis leading to ossification (Joyce et al. 1990) and in a single injection of TGF-β will induce closure of nonhealing skull defects (Beck et al. 1991).

The best characterized member of the TGF-β superfamily is TGF-β1. The name for TGF-β was derived from its ability to transform non-neoplastic fibroblasts in the presence of epidermal growth factor as determined by colony formation in soft agar (Moses et al. 1981; Roberts et al. 1981). Since then, TGF-β has been shown to be a multifunctional cytokine whose major function appears to be promotion of matrix formation and inhibition of cell growth (Moses et al. 1990; Penttinen et al. 1988). Many cells produce TGF-β and most cells have receptors from TGF-β (Massagué 1992), therefore production of active TGF-β appears to be tightly regulated. This tight regulation is achieved by the production of TGF-β in a biologically latent form which must be processed and dissociated for active TGF-β to be released (see Fig. 3). Once released, active TGF-β is quickly cleared by binding to receptors or to storage molecules such as decorin (Yamaguchi et al. 1990), to scavenging clearance molecules such as α2-macroglobulin (O'Connor-McCourt and Wakefield 1987) or to molecules such as thrombospondin which maintain TGF-β in a biologically active form (Murphy-Ullrich et al. 1992). At least three cell types have been shown to activate latent TGF-β. Macrophages treated with γ-interferon (Twardzik et al. 1990), osteoclasts treated with retinol (Oreffo et al. 1989), and mesenchymal cells treated with glucocorticoids (Rowley 1992) will activate latent TGF-β. Endothelial cells cocultured with pericytes will produce active TGF-β (Antonelli-Orlidge et al. 1989). Keratinocytes treated with retinoic acid will secrete active TGF-β2 (Glick et al. 1989) and fetal fibroblasts treated with antiestrogens will also produce active TGF-β (Colletta et al. 1990).

Several latent TGF-β complexes produced by different cell types have been described. Latent TGF-β produced by fibroblasts consists of mature (25-kDa) TGF-β noncovalently associated with a 75- to 80-kDa portion of the precursor (or latency-associated peptide), which in turn is linked by a disulfide linkage to a latent TGF-β binding protein (LTBP) of 190 kDa (Kanzaki et al. 1990) (see Fig. 2). Latent TGF-β produced by platelets is essentially identical, except that the LTBP is approximately 130 kDa and appears to be a proteolytically processed

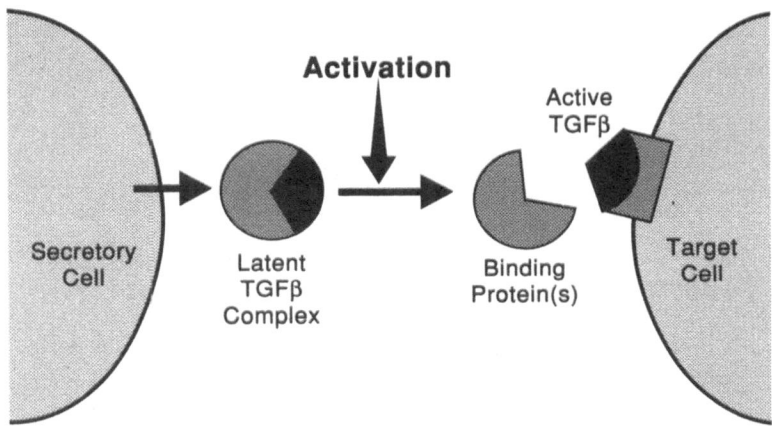

Fig. 3. Latent TGF-β is secreted by cells as one of the latent complexes described in Fig. 2. Mature TGF-β must be released from this complex to be biologically active. Mature or active TGF-β can bind to cellular receptors, can rebind to the latency-associated peptide or can bind to molecules such as α_2-macroglobulin, decorin, or thrombospondin which can inactivate or maintain TGF-β in an active form

form (Miyazono et al. 1988; Wakefield et al. 1988; Kanzaki et al. 1990; Tsuji et al. 1990). LTBP has no covalent linkage with mature TGF-β and is not necessary to confer latency to the complex (Gentry et al. 1987; Bonewald et al. 1991; Kanzaki et al. 1990). At present the function of LTBP is unknown. We have previously found that TGF-β in conditioned medium from bone organ cultures exists in several latent forms (Bonewald et al. 1991). The major form is a 100-kDa precursor complex which lacks LTBP and appears identical to a 100-kDa recombinant form of latent TGF-β expressed in CHO cells (Gentry et al. 1987). This 100-kDa complex has not been reported as a naturally occurring form in such large amounts in other cell systems, suggesting an important function in bone.

Characterization of latent TGF-β complexes in osteoblast-like cells revealed that they produce at least three secreted forms of latent TGF-β (Dallas et al. 1993). MG63 cells produced predominantly the 290-kDa latent TGF-β complexes containing TGF-β1 and TGF-β2, while

Effect of Estrogen ?

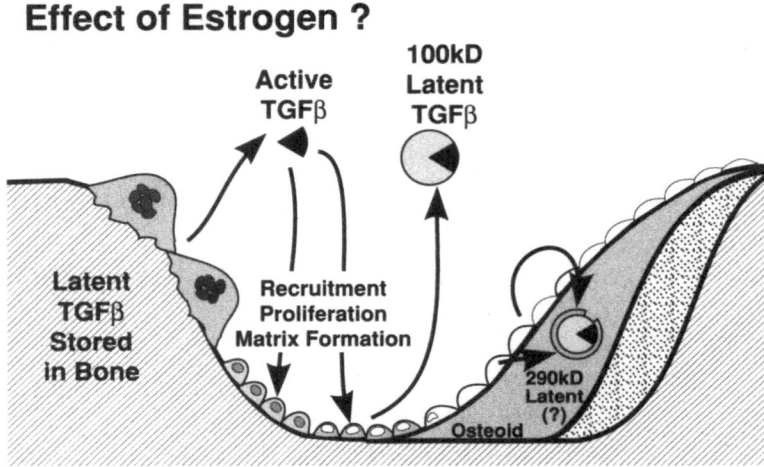

Fig. 4. Expression of TGF-β during the bone remodeling sequence. TGF-β stored as a latent form in bone. Resorbing osteoclasts release active TGF-β, which may inhibit further osteoclastic bone resorption and initiate the cascade of events leading to new bone formation. Active TGF-β is chemotactic and initiates proliferation in osteoblast precursors and is a potent stimulator of matrix formation. These proliferating cells appear to be producing essentially the 100-kDa latent TGF-β complex. This form may be a circulating form, more easily activated, which could have autocrine effects. The more differentiated osteoblasts appear to produce the 290-kDa latent complex containing the LTBP. This complex may be targeted for storage in bone

ROS 17/2.8 cells produced both the 290-kDa and 100-kDa TGF-β1 forms. UMR-106 cells produced almost exclusively the 100-kDa TGF-β1 complex lacking LTBP. These data compliment our previous findings using organ cultures of bone (Bonewald et al. 1991), where three different latent TGF-β complexes were observed. The major form was a 100-kDa complex containing TGF-β1. A higher molecular weight form contained TGF-β1 and a third high molecular weight form contained TGF-β2. These findings also agree with reports showing that the major TGF-β isoform purified from bone is TGF-β1 (200 μg/kg bone) with TGF-β2 present at one tenth this concentration (Seyedin et al. 1985, 1986).

TGF-β is stored in bone in a latent form where it is released by osteoclastic bone resorption (Pfeilschifter and Mundy 1987; Oreffo et al. 1989). Our present experiments suggest that proliferating and early fetal rat calvarial cells produce almost exclusively the 100-kDa latent form of TGF-β. In the mineralizing fetal rat calvarial cultures first described by Bellows and coworkers (1986), the cells are committed to differentiation. In the early phase the cells are proliferating and producing matrix, in the middle stage nodules are being formed, followed by mineralization in the late stage. We have found that as these cells differentiate a switching occurs between the types of latent TGF-β produced (see Fig. 4). As the early proliferating cells make the 100-kDa form, the nodule-forming cells produce the 290-kDa latent complex and express the LTBP within the nodules as detected by immunohistochemistry. We speculate that the 100-kDa form is a circulatory form that may be more easily activated as these proliferating cells are auto-induced by active TGF-β. The 290-kDa form may be a storage form targeted to bone matrix as the nodule-forming cells produce this form, and the LTBP has been shown to possess characteristics of matrix proteins such as calcium-binding sites (Kanzaki et al. 1990) and homology with the matrix protein fibrillan (Olofsson et al. 1992).

5.8 Similar Functions of Estrogen and TGF-β

Estrogen and TGF-β have been shown to have overlapping functions. Estrogen will inhibit expression of bone-resorbing cytokines also important in functions. Estrogen therapy can suppress elevated IL-1 levels in osteoporotic patients (Pacifici et al. 1987) and estrogen can inhibit tumor necrosis factor release by blood mononuclear cells in postmenopausal women, but not premenopausal women or men (Ralston et al. 1990). 17β-Estradiol inhibits IL-6 production by stromal cells and osteoblasts (Girasole et al. 1990). TGF-β is also immunosuppressive by inhibiting cytokine production (Espevich et al. 1990). Estrogen enhances collagen synthesis (Colvard et al. 1990) as does TGF-β (Pfeilschifter et al. 1987). Both estrogen and TGF-β inhibit osteocalcin production by osteoblast-like cells (Bonewald et al. 1990; Aronow et al. 1990). TGF-β expression is increased during fracture healing (Joyce et al. 1990) as is estrogen receptor mRNA expression

(Bolander et al. 1989). As does TGF-β, estrogen inhibits PTH-stimulated bone resorption (Pilbeam et al. 1989; Pfeilschifter et al. 1987), cytokine production and pit formation by osteoclasts (Ousler et al. 1993; Oreffo et al. 1990).

The effects of estrogen may be partially mediated through the production of TGF-β since estrogen enhances TGF-β production. TGF-β was shown to mediate the effects of estrogen on inhibition of cell growth of an osteoblastic cell line UMR-106 (Gray et al. 1989b). Ovariectomy reduces TGF-β concentration in bone (Finkelman et al. 1992). Therefore estrogen may maintain normal osteoblast function and attenuate osteoclast function by inhibiting cytokine production by immune cells and promoting the production of bone proteins such as TGF-β by osteoblasts which play a role in bone formation.

5.9 Effects of Estrogen and Antiestrogens on TGF-β

It is well known that estrogen will stimulate the production of TGF-β1 mRNA in osteoblast cell lines (Komm et al. 1988) and increase latent TGF-β production in isolated human osteoblast cells (Oursler et al. 1991). We have found that 17β-estradiol will induce the production of latent but not active TGF-β in the osteoblast cell lines MG-63, ROS 17/2.8, UMR-106, and SaOS-2. Tamoxifen and toremifen inhibited or had little effect on latent TGF-β produced by these cells (personal observations). Colletta and coworkers (1990) found that these antiestrogens, tamoxifen and toremifene, induced the production of active, not latent TGF-β in human fetal fibroblasts whereas estrogen induced little active TGF-β. This study demonstrated a biological action of antiestrogens in cells lacking classical estrogen receptors. Therefore the mechanism of action of antiestrogens is independent of the presence of estrogen receptor.

5.10 Future Studies

No studies have been performed to determine the effects of antiestrogens on active TGF-β production in bone cells. Could synergism be observed with combination therapy of estrogens and antiestrogens? In theory, estrogens could induce the production of latent TGF-β and antiestrogens induce production of active TGF-β. It is also unknown as to what effects progestins may have on latent and active TGF-β production by bone cells. These questions remain to be answered by future experiments.

References

Ammann P, Rizzoli R, Bonjour JP (1991) Chronic infusion of IGF-I increases bone mineral density (BMD) evaluated sequentially by dual energy x-ray absorptiometry (DXA) in ovariectomized (OVX) osteopenic rats. J Bone Miner Res 6 [Suppl 1]:S218 (abstract)

Antonelli-Orlidge A, Saunders KB, Smith SR, D'Amore PA (1989) An activated form of transforming growth factor beta is produced by coculture of endothelial cells and pericytes. Proc Natl Acad Sci USA 86:4544–4548

Aronow M, Owen TA, Stein GS, Lian JB (1990) Estrogen inhibition of osteocalcin mRNA expression occurs in cultured rat osteoblasts only after formation of a mineralized extracellular matrix. J Bone Miner Res 5 [Suppl 2]:S273 (abstract)

Bain SD, Jensen E, Celino DL, Bailey MC, Lantry MM, Edwards MW (1993) High-dose gestagens modulate bone resorption and formation and enhance estrogen-induced endosteal bone formation in the ovariectomized mouse. J Bone Miner Res 8:219–230

Bankson DD, Rifai N, Williams ME, Silverman LM, Gray TK (1989) Biochemical effect of 17-beta-estradiol on UMR-106 cells. Bone Miner 6:55–63

Barbagallo M, Carbognanani E, Palummeri E, Chiavarini M, Pedrazzoni M, Bracchi PG, Passeri M (1989) The comparative effect of ovarian hormone administration on bone mineral status in oophorectomized rats. Bone 10:113–116

Barengolts EI, Gajardo HF, Rosol TJ, D'Anza JJ, Pena M (1990) Effects of progesterone on post ovariectomy bone loss in aged rats. J Bone Miner Res 5:1143–1147

Baxter RC, Martin JL (1986) Radioimmunoassay of growth hormone-dependent insulin-like growth factor binding protein in human plasma. J Clin Invest 12:1504–1512

Baxter RC, Martin JL (1989) Binding proteins for the insulin-like growth factors: Structure, regulation and function. Prog Growth Factor Res 1:49–68

Beck LS, Deguzman L, Lee WP, Xu Y, McFatridge LA, Gillett NA, Amento EP (1991) TGFβ induces bone closure of skull defects. J Bone Miner Res 6:1257–1265

Bellows CG, Aubin JE, Heersche JNM, Antosz ME (1986) Mineralized bone nodules formed in vitro from enzymatically released rat calvaria populations. Calcif Tissue Int 38:143–154

Blanchard O, Tsagris L, Rappaport R, Duval Beaupere G, Corvol M (1991) Age-dependent responsiveness of rabbit and human cartilage in sex steroids in vitro. J Steroid Biochem Mol Biol 40:711–716

Bolander ME, Joyce ME, Boden SD, Oliver B, Heydemann A (1989) Estrogen receptor mRNA expression during fracture healing the rat detected by polymerase chain reaction amplification. J Bone Miner Res 4 [Suppl 1]:S259 (abstract)

Bonewald LF, Mundy GR (1990) Role of transforming growth factor-beta in bone remodeling. Clin Ortho Rel Res 250:261–276

Bonewald LF, Wakefield L, Oreffo ROC, Escobedo A, Twardzik DR, Mundy GR (1991) Latent forms of transforming growth factors derived from bone cultures: identification of a naturally occurring 100 kDa complex with similarity to recombinant latent TGFβ. Mol Endocrinol 5:741–751

Bonewald LF, Kester MB, Schwartz Z, Swain LD, Khare A, Johnson TL, Leach KJ, Boyan BD (1992) Effects of combining transforming growth factor beta (TGFβ) and 1,25 dihydroxyvitamin D₃ on differentiation of a human osteosarcoma (MG-63). J Biol Chem 267:8943–8949

Canalis EM (1980) Effect of insulin-like growth factor I on DNA and protein synthesis in cultured rat calvaria. J Clin Invest 66:709–719

Carrascosa A, Corvol M, Tsagris L, Rappaport R (1981) Biological effect of estradiol on phosphatase activity in rabbit cultured chondrocytes. Pediatr Res 25:1542 (abstract)

Centrella M, McCarthy TL, Canalis E (1991) Current concepts reveiw – transforming growth factor-beta and remodeling of bone. J Bone Joint Surg 73A:1418–1428

Chen TL, Bates RL, Dudley A, Hammonds GR, Amento EP (1991) Bone morphogenetic protein 2b: stimulation of growth and osteogenic phenotype in rat osteoblast-like cells: comparison with TGFβ1. J Bone Miner Res 6:1387–1393

Christiansen C, Riis BJ (1990) 17β-Estradiol and continuous norethisterone: a unique treatment for established osteoporosis in elderly women. J Clin Endocrinol Metab 71:836–841

Christiansen C, Riis BJ, Nilas L, Rodbro P, Deftos L (1985) Uncoupling of bone formation and resorption by combined oestrogen and progestogen therapy in postmenopausal osteoporosis. Lancet 1:800–801

Clemmons DR (1990) Insulin-like growth factor binding proteins. Trends Endocrinol Metab 1:412–417

Colletta AA, Wakefield LM, Howell FV, van Roozendaal KEP, Danielpour D, Ebbs SR, Sporn MB, Baum M (1990) Anti-oestrogens induced the secretion of active transforming growth factor beta from human fetal fibroblasts. Br J Cancer 62:405–409

Colvard D, Keeting P, Scott R, Spelsberg T, Riggs BL (1990) Effects of estrogen on growth and differentiation of normal human osteoblast-like cells in long-term human marrow cultures. Proc Natl Acad Sci USA 85:5683–5687

Dallas SL, Park-Snyder S, Miyazono K, Twardzik D, Mundy GR, Bonewald LF (1993) Characterization and autoregulation of latent TGFβ complexes in osteoblast-like cell lines: production of a latent complex lacking the latent TGFβ-binding protein (LTBP). J Biol Chem (in press)

De Mellow JSM, Baxter RC (1988) Growth hormone-dependent insulin-like growth factor (IGF) binding protein both inhibits and potentiates IGF-I-stimulated DNA synthesis in human skin fibroblasts. Biochem Biophys Res Commun 156:199–204

Donahue LR, Hunter SJ, Sherblom AP, Rosen C (1990) Age-related changes in serum insulin-like growth factor-binding proteins in women. J Clin Endocrinol Metab 71:575–579

Eriksen EF, Colvard DS, Berg NJ, Graham ML, Mann KG, Spelsberg TC, Riggs BL (1988) Evidence of estrogen receptors in normal human osteoblast-like cells. Science 241:84–86

Ernst M, Schmid C, Froesch ER (1987) 17-beta-estradiol stimulates proliferation and type I procollagen gene expression in primary osteoblasts. In: Osteoporosis. Christiansen C, Johansen J, Riis BJ (eds) Osteopress, Copenhagen, pp 198–201

Ernst M, Health JK, Rodan GA (1989a) Estradiol enhances insulin-like growth factor I mRNA and has anabolic effects in osteoblastic cells from rat calvariae and long bones. J Bone Miner Res 4 [Suppl 1]:S256 (abstract)

Ernst M, Schmid C, Froesch ER (1989b) Phenol red mimics biological action of estradiol: enhancement of osteoblast proliferation in vitro and type I collagen gene expression in bone and uterus of rats in vivo. J Steroid Biochem 33:907–914

Espevich T, Waage A, Faxvaag A, Shalaby MR (1990) Regulation of interleukin-2–6 production from T-cells: Involvement of interleukin-1 and transforming growth factor-β. Cell Immunol 126:47–56

Farley JR, Tarbaux N, Murphy LA, Masuda T, Baylink DJ (1987) In vitro evidence that bone formation may be coupled to resorption by release of mitogens(s) from resorbing bone. Metabolism 36:314–321

Fentiman IS, Caleffi M, Rodin A, Murby B, Fogelman I (1989) Bone mineral content of women receiving tamoxifen for mastalgia. Br J Cancer 60:262–264

Finkelman RD, Eason AL, Bell NH, Baylink DJ (1992) Ovariectomy selectively reduces the concentration of transforming growth factor beta in rat bone. Proc Natl Acad Sci USA 89:12190–12193

Fornander T, Rutqvist LE, Sjoberg HE, Blomqvist L, Mattsoson A, Glas U (1990) Long-term adjuvant tamoxifen in early breast cancer: effect on bone mineral in postmenopausal women. J Clin Oncol 8:1019–1024

Froger-Gaillard B, Hossenlopp P, Adolphe M, Binoux M (1989) Production of insulin-like growth factor and their binding proteins by rabbit articular chondrocytes: relationships with cell multiplications. Endocrinology 124:2365–2372

Gallagher JC, Wilkinson R (1973) The effect of ethinyloestradiol on calcium and phosphorus metabolism of postmenopausal women with primary hyperparathyroidism. Clin Sci Mol Med 45:785–802

Gentry LE, Webb NR, Lim JG, Brunner AM, Ranchalis JE, Twardzik DR, Lioubin MN, Marquardt-Purchio AF (1987) Type 1 transforming growth factor beta: amplified expression and secretion of mature and precursor polypeptides in Chinese hamster ovary cells. Mol Cell Biol 7:3418–3427

Girasole G, Sakagami Y, Hustmyer FB, Yu XP, Derrigs HG, Boswell S, Peacock M, Boder G, Manolagas SC (1990) 17-β estradiol inhibits cytokine induced IL-6 production by bone marrow stromal cells and osteoblasts. J Bone Miner Res 5 [Suppl 2]:S273 (abstract)

Girasole G, Jilka RL, Passeri G, Boswell S, Boder G, Williams DC, Manolagas SC (1992) 17-β Estradiol inhibits interleukin-6 production by bone marrow-derived stromal cells and osteoblasts in vitro. A potential mechanism for the antiosteoporotic effect of estrogens. J Clin Invest 89:883–891

Glick AB, Flanders KC, Danielpour D, Yuspa SH, Sporn MB (1989) Retinoic acid induces active transforming growth factor-β in cultured keratinocytes and mouse epidermis. Cell Regulation 1:87–97

Gray TK, Flynn TC, Gray KM, Nabell LM (1987a) 17-beta-estradiol acts directly on the clonal osteoblastic cell line UMR-106. Proc Natl Acad USA 84:6267–6271

Gray TK, Korach K, Nabell LM, Flynn TC, Gray KM, Dodd RC, Sivam G, Williams ME, Cohen MS (1987b) Mechanisms of estradiol's actions on bone. In: Osteoporosis. Christiansen C, Johansen J, Riis BJ (eds) Osteopress, Copenhagen, pp 513–526

Gray TK, Lipes B, Linkhart T, Mohan S, Baylink D (1989a) Transforming growth factor beta mediates the estrogen induced inhibition of UMR-106 cell growth. Connect Tiss Res 20:23–32

Gray TK, Mohan S, Linkhart TA, Baylink DJ (1989b) Estradiol stimulates in vitro secretion of insulin-like growth factors by the clonal osteoblastic cell line UMR-106. Biochem Biophys Res Commun 158:407–412

Jennings JC, Mohan S (1990) Heterogeneity of latent-transforming growth factor-beta isolated from bone matrix proteins. Endocrinology 126:1014–1021

Jilka RL, Hangoc G, Girasole G, Passeri G, Williams DC, Abrams JS, Boyce B, Broxmeyer H, Manologas SC (1992) Increased osteoclast development after estrogen loss: mediation by interleukin-6. Science 25:88–91

Jordan VC, Phelps E, Lindgren JU (1987) Effects of anti-estrogens on bone in castrated and intact female rats. Breast Cancer Res Treat 10:31–35

Joyce ME, Jingushi S, Bolander ME (1990) Transforming growth factor-beta in the regulation of fracture repair. Orthop Clin North Am 21:199–209

Joyce ME, Roberts AB, Sporn MB, Bolander ME (1990) Transforming growth factor β and the initiation of chondrogenesis and osteogenesis in the rat femur. J Cell Biol 110:2195–2207

Kalu DN, Liu CC, Salerno E, Salih M, Echon R, Ray M, Hollis BW (1991) Insulin-like growth factor-I partially prevents ovariectomy-induced bone loss: a comparative study with human parathyroid hormone (1–38). J Bone Miner Res 6:S221 (abstract)

Kanzaki T, Olofsson A, Moren A, Wernstedt C, Hellman U, Miyazono K, Welsh CL, Heldin CH (1990) TGFβ1 binding protein: a component of the large latent complex of TGFβ1 with multiple repeat sequences. Cell 61:1051–1061

Keeting P, Scott R, Colvard D, Han IK, Spelsberg TC, Riggs BL (1991) Lack of a direct effect of estrogen on proliferation and differentiation of normal human osteoblast-like cells. J Bone Miner Res 6:297-304

Kerhl H (1990) Pharmacokinetics of oestrogens and progestogens. Maturitas 12:171–197

Komm BS, Terpening CM, Benz DJ, Graeme KA, Gallegos A, Korc M, Greene GL, O'Malley BW, Haussler MR (1988) Estrogen binding, receptor mRNA, and biologic response in osteoblast-like osteosarcoma cells. Science 241:81–84

LaTour D, Mohan S, Linkhart TA, Baylink DJ, Strong DD (1990) Inhibitory insulin-like growth factor-binding protein: cloning, complete sequence, and physiological regulation. Mol Endocrinol 4:1806–1814

Liel Y, Kraus S, Levy J, Shany S (1992) Evidence that estrogens modulate activity and increase the number of 1,25-dihydroxyvitamin D receptors in osteoblast-like cells (ROS 17/2.8). Endocrinology 130:2597–2601

Lindahl A, Isgaard J, Isaksson OGP (1987) Growth hormone in vivo potentiates the stimulatory effect of insulin-like growth factor-I in vitro on colony formation of epiphyseal chondrocytes isolatd from hypophysectomized rats. Endocrinology 121:1070–1075

Lindsay R, Hart DM, Purdie D, Ferguson MM, Clark AS, Kraszewski A (1978) Comparative effects of oestrogen and progestogen on bone loss in postmenopausal women. Clin Sci Mol Med 54:193–195

Linkhart TA, Mohan S (1989) Parathyroid hormone stimulates release of insulin-like growth factor-I (IGF-I) and IGF-II from neonatal mouse calvaria in organ culture. Endocrinology 125:1484–1491

Linkhart TA, Jennings JC, Mohan S, Wakely GK, Baylink DJ (1986) Characterization of mitogenic activities extracted from bone matrix. Bone 7:479–487

Love RR, Mazess RB, Barden HS, Epstein S, Newcomb PA, Jordan VC, Carbone PP, Demets DL (1988) Effects of tamoxifen on bone mineal density in postmenopausal women with breast cancer. J Natl Cancer Inst 81:1086–1088

Love RR, Mazess RB, Tormey DC, Barden HS, Newcomb PA, Jordan VC (1992) Bone mineral density in women with breast cancer treated with adjuvant tamoxifen for at least two years. Breast Cancer Res Treat 12:297–301

Luyten FP, Cunningham NS, Ma S, Muthukumaran N, Hammonds Jr RG, Nevins WB, Wood WI, Reddi AH (1989) Purification and partial amino acid sequence of osteogenin, a protein initiating bone differentiation. J Biol Chem 264:13377–13380

Marcelli C, Yates AJP, Mundy GR (1990) In vivo effects of human recombinant transforming growth factor beta on bone turnover in normal mice. J Bone Miner Res 5:1087–1096

Massagué J (1992) Receptors for the TGF-beta family. Cell 69:1067–1070

McCusker RH, Campion DR, Jones WK, Clemmons DR (1989) The insulin-like growth factor I (IGF-I)-binding protein complex is a better mitogen than free IGF-I. Endocrinology 125:766–772

Miyazono K, Hellman U, Werstedt C, Heldin CH (1988) Latent high molecular weight complex of transforming growth factor β1. Purification from human platelets and structural characterization. J Biol Chem 263:6407–6415

Mohan S, Linkhart T, Jennings J, Baylink D (1988) Primary structure of human skeletal growth factor: homology with human insulin-like growth factor II. Biochim Biophys Acta 966:44–45

Moses HL, Branum EL, Proper JA, Roberson RA (1981) Transforming growth factor production by chemically transformed cells. Cancer Res 41:2842–2848

Moses HL, Yang EY, Pietenpol JA (1990) TGFβ stimulation and inhibition of cell proliferation: new mechanistic insights. Cell 63:245–247

Mueller K, Cortesi R (1991) Insulin-like growth factor-I increases trabecular bone mass in ovariectomized rat. J Bone Miner Res 6 [Suppl 1]:S221 (abstract)

Murphy-Ullrich JE, Schultz-Cherry S, Hook M (1992) Transforming growth factor-complexes with thrombospondin. Mol Biol Cell 3:181–188

Nasatsky E, Schwartz Z, Boyan BD, Soskolne WA, Ornoy A (1993) Sex-dependent effects of 17-beta-estradiol on chondrocyte differentiation in culture. J Cell Physiol 154:359–367

Nilsson A, Carlsson B, Isgaard J, Isaksson OGP, Rymo L (1990) Regulation by GH of insulin-like growth factor-I mRNA expression in rat epiphyseal growth plate as studied with in-situ hybridization. J Endocrinol 125:67–74

Noda M, Camilliere JJ (1989) In vivo stimulation of bone formation by transforming growth factor β. Endocrinology 124:2991–2994

O'Connor-McCourt MD, Wakefield LM (1987) Latent transforming growth factor beta in serum: a specific complex with alpha 2 macroglobulin. J Biol Chem 262:14090–14099

Olofsson A, Miyazono K, Kanzaki T, Colosetti P, Engstrom U, Heldin CH (1992) Transforming growth factor-β1, -β2 and -β3 secreted by a human glioblastoma cell line: Identification of small and different forms of large latent complexes. J Biol Chem 267:19482–19488

Oreffo ROC, Mundy GR, Seyedin SM, Bonewald LF (1989) Activation of the bone derived latent TGFβ complex by isolated osteoclasts. Biochem Biophys Res Commun 158:817–823

Oreffo ROC, Bonewald L, Kukita A, Garrett IR, Seyedin SM, Rosen D, Mundy GR (1990) Inhibitory effects of the bone-derived growth factors osteoinductive factor and transforming growth factor β on isolated osteoclasts. Endocrinology 126:3069–3075

Oursler MJ, Cortese C, Keeting P, Anderson MA, Bonde SK, Riggs BL, Spelsberg TC (1991) Modulation of transforming growth factor-beta production in normal human osteoblast-like cells by 17 beta-estradiol and parathyroid hormone. Endocrinology 129:3313–3320

Oursler MJ, Pederson L, Pyfferoen J, Osdoby Ph, Fitzpatrick L, Spelsberg TC (1993) Estrogen modulation of avian osteoclast lysosomal gene expression. Endocrinology 132:1373–1380

Ozkaynak E, Rueger DC, Drier EA, Corbett C, Ridge RJ (1990) OP-1 cDNA encodes an osteogenic protein in the TGF-beta family. EMBO J 9:2085–2093

Pacifici R, Rifas L, Teitelbaum S, Slatopolsky E, McCracken R, Bergfeld M, Lee W, Avioli LV, Peck WA (1987) Spontaneous release of interleukin-1 from human blood monocytes reflects bone formation in idiopathic osteoporosis. Proc Natl Acad Sci USA 84:4616–4620

Pacifici R, Rifas L, McCracken R, Vered I, McMurtry C, Avioli LV, Peck WA (1989) Ovarian steroid treatment blocks a postmenopausal increase in blood monocyte interleukin-1 release. Proc Natl Acad Sci USA 86:2398–2402

Penttinen RP, Kobayashi S, Bornstein P (1988) Transforming growth factor β increases mRNA for matrix proteins both in the presence and in the absence of changes in mRNA stability. Proc Natl Acad Sci USA 85:1105–1108

Pfeilschifter J, Mundy GR (1987) Modulation of transforming growth factor beta activity in bone cultures by osteotropic hormones. Proc Natl Acad Sci USA 84:2024–2028

Pfeilschifter J, D'Souza SM, Mundy GR (1987) Effects of transforming growth factor-beta on osteoblastic osteosarcoma cells. Endocrinology 121:212–218

Pfeilschifter J, Seyedin SM, Mundy GR (1988) Transforming growth factor beta inhibits bone resorption in fetal rat long bone cultures. J Clin Invest 82:680–685

Pilbeam CC, Klein-Nulend J, Raisz LG (1989) Inhibition by 17-beta-estradiol of PTH stimulated resorption and prostaglandin production in cultured neonatal mouse calvaria. Biochem Biophys Res Commun 163:1319–1324

Pioli G, Basini G, Pedrazzoni M, Musetti G, Ulietti V, Bresciani D, Villa P, Bacchi A, Hughes D, Russell G, Passeri M (1992) Spontaneous release of interleukin-1 and interleukin-6 by peripheral blood mononuclear cells after oophorectomy. Clin Sci 83:503–507

Purohit A, Flanagan AM, Reed MJ (1992) Estrogen synthesis by osteoblast cell lines. Endocrinology 131:2027–2029

Ralston ST, Graham R, Russell G, Maxine G (1990) Estrogen inhibits release of tumor necrosis factor from peripheral blood mononuclear cells in postmenopausal women. J Bone Miner Res 5:983–988

Rawlings ND, Barrett AJ (1990) Evolutionary families of peptidases. Biochem J 266:622–624

Reed MJ, Coldham NG, Patel SR, Ghilchik MW, James VHT (1992) Interleukin-1 and interleukin-6 in breast cyst fluid: their role in regulating aromatase activity in breast cancer cells. J Endocrinol 132:R5

Rickard DJ, Gowen M, MacDonald BR (1993) Proliferative responses to estradiol, IL-1α and TGFβ by cells expressing alkaline phosphatase in human osteoblast-like cell cultures. Calcif Tissue Int 52:227–233

Roberts AB, Angno MA, Lamb LC, Smith JM, Sporn MD (1981) New class of transforming growth factors potentiated by epidermal growth factor: isolation from non-neoplastic tissues. Proc Natl Acad Sci USA 78:5339

Rosenfeld RG, Lamson G, Pham H, Oh Y, Conover C, De Leon DD, Donovan SM, Ocrant I, Giudice L (1990) Insulin-like growth factor-binding proteins. Recent Prog Horm Res 46:99–159

Rosner IA, Malmeud CJ, Hassid AI, Goldberg VM, Boja BA, Moskowitz RW (1983) Estradiol and tamoxifen stimulation of lapine articular chondrocyte prostaglandin synthesis. Prostaglandins 26:123–138

Rowley DR (1992) Glucocorticoregulation of transforming growth factor-β activation in urogenital sinus mesenchymal cells. Endocrinology 131:471–478

Ryan WG, Wolter J, Bagdade JD (1991) Apparent beneficial effects of tamoxifen on bone mineral content in patients with breast cancer. Osteoporosis Int 2:39–41

Scharla SH, Strong DD, Mohan S, Baylink DJ, Linkhart TA (1991) 1,25-Di-hydroxyvitamin D$_3$ differentially regulates the production of insulin-like growth factor I (IGF-I) and IGF-binding protein-4 in mouse osteoblasts. Endocrinology 129:3139–3146

Schmid C, Ernst M, Zapf J, Froesch ER (1989) Release of insulin-like growth factor carrier proteins by osteoblasts: stimulation by estradiol and growth hormone. Biochem Biophys Res Commun 160:788–794

Schmidt C, Steiner R, Froesch EF (1983) Insulin-like growth factors stimulate synthesis of nucleic acids and glycogen in cultured calvaria cells. Calcif Tissue Int 35:578–585

Schwartz Z, Soskone WA, Neubauer T, Goldstein M, Adi S, Ornoy A (1991) Direct and sex-specific enhancement of bone formation and calcification by sex steroids in fetal mice long bone in vitro (biochemical and morphometric study). Endocrinology 129:1167–1174

Scranton PE, McMaster JH, Diamond PE (1975) Hormone suppression of DNA synthesis in cultured chondrocytes and osteosarcoma cell lines. Clin Orthop Rel Res 112:340–348

Selby PL, Peacock M (1986) Ethinyl estradiol and norethindrone in the treatment of primary hyperparathyroidism in postmenopausal women. N Engl J Med 317:1481–1485

Seyedin SM, Thomas TC, Thompson AY, Rosen DM, Piez KA (1985) Purification and characterization of two cartilage-inducing factors from bovine demineralized bone. Proc Natl Acad Sci USA 82:2267–2271

Seyedin SM, Thompson AY, Bentz H, Rosen DM, McPherson JM, Conti A, Siegel NR, Gallupi GR, Piez KA (1986) Cartilage-inducing factor-A. Apparent identity to transforming growth factor-β. J Biol Chem 261:5693–5695

Singh A, Reed MJ (1991) Insulin-like growth factor type I and insulin-like growth factor type II stimulate oestradiol 17β-hydroxysteroid dehydrogenase (reductive) activity in breast cancer cells. J Endocrinol 129:R5

Spencer EM, Liu CC, Si ECC, Howard GA (1991) In vivo actions of insulin-like growth factor-I (IGF-I) on bone formation and resorption in rats. Bone 12:21–26

Stock JL, Coderre JA, McDonald B, Rosenwasser LJ (1989) Effects of estrogen in vivo and in vitro on spontaneous interleukin-1 release by monocytes from postmenopausal women. J Clin Endocrinol Metab 68:364–368

Takahashi MM, Noumura T (1987) Sexually dimorphic and laterally asymmetric development of the embryonic duck syrinx: effect of estrogen on in vitro cell proliferation and chondrogenesis. Develop Biol 121:417–422

Takano-Yamamoto TT, Rodan G (1990) Direct effects of 17β-estradiol on trabecular bone in ovariectomized rats. Proc Natl Acad Sci USA 87:2172–2176

Tarken S, Siris E, Seldin D, Flaster E, Hyman G, Lindsay R (1989) Effects of tamoxifen on spinal bone density in women with breast cancer. J Natl Cancer Inst 81:1086–1088

Tsuji T, Okada F, Yamaguchi K, Nakamura T (1990) Molecular cloning of the large subunit of transforming growth factor type B masking protein and expression of mRNA in various rat tissues. Proc Natl Acad Sci USA 87:8835–8839

Turner RT, Wakley GK, Hannon KS, Bell NH (1987) Tamoxifen prevents the skeletal effects of ovarian hormone deficiency in rats. J Bone Miner Res 2:449–456

Turner RT, Wakley GK, Hannon KS, Bell NH (1988) Tamoxifen inhibits osteoclast-mediated resorption of trabecular bone in ovarian hormone-deficient rats. Endocrinology 122:1146–1150

Twardzik DR, Mikovits JA, Ranchalis JE, Purchio AF, Ellingsworth L, Ruscetti FW (1990) γ-Interferon activation of latent transforming growth factor beta by human monocytes. Ann NY Acad Sci 593:276–284

Urist MR (1965) Bone formation by autoinduction. Science 150:893–899

Wakefield LM, Smith DM, Flanders KC, Sporn MB (1988) Latent transforming growth factor-beta from human platelets. A high molecular weight complex containing precursor sequences. J Biol Chem 263:7646

Wergedal JE, Mohan S, Taylor AK, Baylink DJ (1986) Skeletal growth factor is produced by human osteoblast-like cells in culture. Biochim Biophys Acta 889:163–170

Wergedal JE, Mohan S, Lundy M, Baylink DJ (1990) Skeletal growth factor and other growth factors known to be present in bone matrix stimulate proliferation and protein synthesis in human bone cells. J Bone Miner Res 5:179–186

Wozney JM (1992) The bone morphogenetic protein family and osteogenesis. Mol Reproduction Dev 32:160–167

Wozney JM, Rosen V, Celeste AJ, Mitsock LM, Whitters MJ, Kriz RW, Hewick RM, Wang EA (1988) Novel regulators of bone formation: Molecular clones and activities. Science 242:1528–1534

Yamaguchi Y, Mann DM, Ruoslahti E (1990) Negative regulation of transforming growth factor β by the proteoglycan decorin. Nature 346:261–284

Yamaguchi A, Katagiri T, Ikeda T, Wozney JM, Rosen V, Wang EA, Kahn AJ, Suda T, Yoshiki S (1991) Recombinant human bone morphogenetic protein 2 stimulates osteoblastic maturation and inhibits myogenic differentiation in vitro. J Cell Biol 113:681–687

6 Estrogens, Cytokines, and Bone Metabolism

Stavros C. Manolagas

6.1 Introduction

Remodeling of bone is a process that goes on continuously throughout adult life. This is accomplished by the resorption of old bone by osteoclasts and the subsequent formation of new bone by osteoblasts [82]. These two events are tightly coupled and are responsible for the renewal of the skeleton while maintaining its anatomical and structural integrity. Under normal physiologic circumstances, bone remodeling proceeds in highly regulated cycles in which osteoclasts adhere to bone and subsequently remove it by acidification and proteolytic digestion. After osteoclasts have left the resorption site, osteoblasts invade the area and begin the process of new bone formation by secreting osteoid (a matrix of collagen and other proteins), which is eventually mineralized into new bone.

An imbalance of bone remodeling in favor of resorption is the cardinal feature of several disease states of the skeleton, including the osteoporosis that ensues upon loss of ovarian function. Indeed, reduction of skeletal mass caused by an imbalance of bone resorption over bone formation represents the hallmark of postmenopausal osteoporosis. The deficit of bone in osteoporosis exists primarily where bone is in contact with the bone marrow, namely, in the trabecular bone. In fact, the rapid loss of trabecular bone, precipitated by the loss of ovarian function, is associated with perforation of individual trabecular plates caused by increased frequency of remodeling activity and an increase in the number of osteoclasts present in trabecular bone [64,117] as well as an increase in osteoclastic resorption depth [83]. Thus, loss of estrogens appears to cause an imbalance in the normally coupled bone resorption and bone formation processes, due to the combination of increased bone resorption and inadequate compensation by bone formation [23].

During the last decade, there have been major advances in our understanding of the ontogeny of osteoclasts and osteoblasts, the interplay between them, and the systemic and local factors that regulate their development and the coupling of their function. This information has provided a much clearer picture of bone metabolism and new insights into the pathophysiology of postmenopausal osteoporosis.

In this article, I review evidence which provides a mechanistic explanation for the protective effect of estrogens on bone homeostasis; and points to interleukin-6 (IL-6) as a critical pathogenetic factor in the bone loss caused by estrogen deficiency.

6.2 The Relationship Between Bone Marrow and Bone

It is now well established that both osteoclasts and osteoblasts are derived from precursors occurring in the bone marrow; and that besides their anatomical juxtaposition, the bone marrow and bone interact extensively and are critical for each other's function.

The intersinusoidal spaces of the bone marrow are filled with a variety of cell types, including fibroblastoid, adventitial reticular, adipocytic, macrophagic, endothelial cells, as well as cells with osteogenic potential that are collectively referred to as the stromal tissue [18,121]. The mesenchymal fibroblastic and reticular cells have extensive cytoplasmic extensions that make intimate contact with hematopoietic progenitors as well as with cellular processes of reticular cells present in the nearby intersinusoidal spaces. These cells are generally alkaline phosphatase positive and are considered the major cell type that regulates hematopoiesis.

Bone marrow stromal cells are essential for the formation of various types of blood cells such as lymphocytes, granulocytes, platelets, and erythrocytes, as well as monocytes, macrophages and osteoclasts [72,103]. These specialized cell types appear to derive from self-renewing totipotential cells [21]. The formation of fully differentiated cells from a single progenitor involves the progression from a state of less-differentiated, non-lineage-specific cells to a state of greater differentiation and highly restricted lineage specificity. This process is controlled by cytokines and growth factors that are synthesized and released locally by stromal cells of the bone marrow.

It is now widely accepted that osteoclasts are derived from progenitors that originate from hematopoietic cells of the bone marrow. Specifically, osteoclasts develop from cloned pluripotent stem cells as well as from the multipotential hematopoietic cells colony forming unit-granulocyte/erythrocyte/megakaryocyte/macrophage (CFU-GEMM) and CFU-granulocyte/macrophage (CFU-GM), but not from the unipotential CFU-M and CFU-G [32,35,55,59]. Nonetheless, the point at which the committed osteoclast progenitor in the marrow diverges from the macrophage lineage is not clear. Intriguingly, monocyte/macrophage preparations obtained from tissues such as lung, spleen, and peripheral blood, cultured in the presence of the appropriate supportive stromal cells, can give rise to osteoclasts [118].

Osteoclasts may reach bone either by chemotaxis or via the circulation; the former is probably more relevant to areas of bone that are in direct contact with the marrow, i.e., the trabecular bone. Evidence that osteoclasts can reach bone via the circulation has been provided by the results of parabiosis studies in which two experimental animals are joined along their lateral body wall, allowing the exchange of blood cells between them [68]. Indeed, it has been found in such studies that blood cells from a normal mouse can correct the paucity of functional osteoclasts of an osteopetrotic mouse. Consistent with the evidence that osteoclast progenitors can arise from the blood stream, CFU-GM have been recently identified in peripheral blood [102].

The stromal cells of the bone marrow, as well as osteoblastic cells, also originate from pluripotent stem cells. As opposed to most hematopoietic cells, stromal and osteoblastic cells adhere to plastic surfaces. Stromal cells with osteogenic potential, as well as with the potential to become fibroblasts, chondrocytes, adipocytes, and muscle cells, are formed in marrow cell cultures as adherent colonies. The common progenitor which gives rise to these colonies is termed colony forming unit-fibroblasts (CFU-F) [77]. Cells present in CFU-F colonies can induce both bone formation and hematopoiesis when transferred under the kidney capsule [26]; and bone can be formed in diffusion chambers containing bone marrow cells or CFU-F harvested from cultures of marrow cells [5]. Furthermore, stromal cells can form calcified nodules in vitro in the appropriate milieu [69].

Marrow stromal cells and osteoblastic cells exhibit an extensive overlap in their phenotypic properties. Indeed, bone marrow-derived stromal cell lines express phenotypic markers of osteoblasts such as alkaline phosphatase and collagen type I [8,9]. Conversely, osteoblastic cells are capable of secreting the same colony stimulating factors and cytokines which are secreted by stromal cells such as IL-6, IL-11, granulocyte/macrophage–colony stimulating factor (GM-CSF), and macrophage–colony stimulating factor (M-CSF) [37]. At this stage, it is not known whether the stromal cells responsible for the control of hematopoiesis and the stromal cells which give rise to osteoblasts represent distinct lineages or different maturation stages along the differentiation pathway of a single cell lineage.

6.3 The Role of Systemic and Local Factors in the Regulation of Osteoclast Development

A major advance in our understanding of osteoclast development has been the elucidation of the paramount importance of stromal/osteoblastic cells for osteoclast formation. Indeed, it is now established that the development of osteoclasts from hematopoietic precursors cannot be accomplished unless stromal/osteoblastic cells are present. Moreover, it is established that cells of the stromal/osteoblastic lineage mediate the effects of both systemic hormones and locally produced factors which stimulate osteoclast development [14,109].

The two major hormones of the calcium homeostatic system, namely, parathyroid hormone (PTH) and 1,25-dihydroxyvitamin D_3 [1, 25$(OH)_2D_3$], are potent stimulators of osteoclast formation both in vitro and in vivo [6,30,36,56,74,110]. Nevertheless, although PTH is a potent stimulator of osteoclast formation, PTH receptors have not been detected on fully differentiated mammalian osteoclasts [98,99]; however, such receptors have been demonstrated in avian osteoclasts [1]. Whether this represents a real difference in the expression of this receptor among species, as opposed to an apparent difference caused by methodologic reasons, remains unclear. In any event, in vitro addition of PTH to cultures of fully differentiated mammalian osteoclasts does not affect osteoclast motility nor does it stimulate the ability of these cells to resorb bone. However, when mammalian osteoclasts are cultured in the presence of osteoblastic cells, increased osteoclastic bone resorption can be induced by the addition of PTH to these cultures [70]. In addition, PTH can stimulate osteoclast development from hematopoietic progenitors such as those found in bone marrow cultures [110]. However, similar to evidence from studies with fully differentiated osteoclasts, this effect requires the presence of stromal/osteoblastic cells [2,125]. Consistent with the evidence that PTH regulates either osteoclast development or the activity of fully differentiated osteoclasts via its actions on stromal/osteoblastic cells, PTH receptors have been found on both adventitial reticular cells of the marrow stroma and mature, bone-forming osteoblasts [98,99].

1,25$(OH)_2D_3$ stimulates the fusion and differentiation of hematopoietic precursors of osteoclasts into mature osteoclasts [110]. This property, however, is not specific for cells of the osteoclast lineage since 1,25$(OH)_2D_3$ can also cause fusion (as well as promote the differentiation) of peripheral blood monocytes [108]. Similar to the case with PTH, 1,25$(OH)_2D_3$ stimulates the activity of differentiated mammalian osteoclasts. This effect, however, requires the presence of stromal/osteoblastic cells [71]. This, and the observation that stromal cells express the receptor for, and respond to, 1,25$(OH)_2D_3$, indicate that the prodifferentiating effects of 1,25$(OH)_2D_3$ on hematopoietic progenitors alone may not account for the effects of the hormone on os-

teoclastogenesis [7]. Hence, additional actions on the cells of stro-
mal/osteoblastic lineage are involved in the osteoclastogenic effects of
this hormone.

Besides the systemic hormones PTH and $1,25(OH)_2D_3$, an array of
factors produced locally in the bone microenvironment can stimulate
osteoclast development. Since the early stages of hematopoiesis and
osteoclastogenesis proceed along identical pathways, it is not surprising that this group of factors includes the same cytokines and CSFs
that are involved in hematopoiesis. To date, the list of cytokines and
CSFs implicated in the regulation of osteoclast development includes
the interleukins IL-1, IL-3, IL-6, IL-11, and tumor necrosis factor
(TNF), GM-CSF, M-CSF, and the leukemia inhibitory factor (LIF).

IL-1 and TNF represent the original paradigms of cytokines that
can influence the bone resorption process. IL-1 refers to two proteins
(IL-1α and IL-1β) that display a wide spectrum of properties, including actions on the immunologic and hematopoietic systems [19]. TNF
also exists in two forms, namely TNF-α and TNF-β [119]. IL-1, TNF,
and IL-6 are interacting cytokines, the induction of which is orches-
trated by a variety of stimuli including the cytokines themselves [3]. In
addition, some IL-1- and TNF-induced responses, such as production
of acute phase proteins, cachexia, anemia, and human chorionic gonadotropin production by trophoblasts are mediated by IL-6 [61,75,106].

Administration of IL-1 or TNF to rodents causes increased bone remodeling [13,50]. In addition, it has been demonstrated that IL-1 and
TNF stimulate osteoclast development from hematopoietic progenitors
in vitro [88]. Further, IL-1 and TNF can stimulate the release of calcium from bone explants [11,28]. The effects of IL-1 and TNF on osteoclastic cells, however, seem to be mediated indirectly through the
actions of these cytokines on stromal/osteoblastic cells [95,113,114].
Indeed, neither of these agents can stimulate isolated osteoclasts to ex-
cavate pits from the surface of smooth cortical bone slices. TNF is
100-fold less potent than IL-1 in inducing bone resorption, but TNF
strongly synergizes with IL-1 to promote this process [107].

Several studies have suggested that osteoblasts isolated from adult
human trabecular bone, neonatal murine calvarial cells, or osteoblastic
cell lines are capable of producing IL-1 and TNF constitutively or in
response to stimulation by lipopolysaccharide or other cytokines
[16,29,53,54,94]. It is important to note, however, that the production

of IL-1 and TNF by these cells appears to be very small (in the 1–10 pM range) and it is not regulated by the systemic hormones PTH or 1,25(OH)$_2$D$_3$. Nevertheless, it appears that despite the very low level of production of TNF in the bone microenvironment, this particular cytokine may serve as an important amplifier for the effects of other local cytokines and systemic hormones that stimulate osteoclast development [47].

During the last 5 years, IL-6 has attracted major attention among the cytokines that are relevant to bone. Indeed, it has been established that IL-6, besides its important role in the early stages of hematopoiesis, can also stimulate osteoclastogenesis. Thus, IL-6 synergizes with IL-3 to stimulate the development of CFU-GM [43,124] and stimulates early osteoclast precursor formation from cells present in CFU-GM colonies [60]. IL-6 also enhances osteoclast formation and bone resorption in cultures of fetal mouse bone in vitro [44,65] and stimulates bone resorption cooperatively with IL-1 in vivo [12]. However, IL-6 does not appear to affect bone resorption in cultured neonatal calvariae [4], which contain a predominance of postmitotic osteoclast progenitors and relatively few of the primitive myeloid progenitors which are the principal targets for IL-6.

Unlike IL-1 and TNF, which are produced in picomolar concentrations, IL-6 is produced in nanomolar quantities by both stromal cells and osteoblastic cells in response to stimulation by both locally produced cytokines, such as IL-1 and TNF [16,27,44,62], growth factors, such as TGF-β [41], and systemic hormones, such as PTH and 1,25(OH)$_2$D$_3$ [25,47,63,65]. Interestingly, among bone marrow-derived stromal cells there is a phenotype-related specificity in the IL-6 response to PTH as a result of a phenotype-specific expression of the PTH receptor. Furthermore, there is evidence that the neuropeptide calcitonin gene-related peptide (CGRP) regulates IL-6 production by bone marrow stromal cells [100]. This raises the possibility that IL-6 might be relevant to an interplay between neuronal stimuli and bone metabolism. Receptors for IL-6 have recently been demonstrated in human osteoclastoma cells, and IL-6 was shown to stimulate the resorptive activity of these cells [76], suggesting that IL-6 may affect not only the early stages of osteoclast development but also mature osteoclasts. However, it is unclear whether normal osteoclasts, as opposed to osteoclastoma cells, also respond to IL-6.

At this stage, the physiologic role of IL-6 in osteoclast development is unclear. In fact, existing evidence suggests that IL-6 may not be involved in osteoclast development under physiologic conditions either because this cytokine is redundant in terms of osteoclast development or its levels in the bone marrow are kept below a critical threshold relative to the sensitivity of the osteoclastogenesis process for IL-6 [49]. On the other hand, IL-6 seems to attain major importance for osteoclastogenesis in pathologic states characterized by the dysregulated production of this cytokine. Indeed, as will be discussed in subsequent sections, IL-6 seems to play a causative role in the pathologic bone resorption associated with the syndrome of postmenopausal osteoporosis [49], as well as multiple myeloma [57] and Paget's disease [97].

IL-11 is the latest member of the group of interleukins involved in osteoclastogenesis. IL-11 was recently cloned from stromal cells [87]. Very recent evidence from the author's laboratory indicates that IL-11 is also produced by osteoblastic cells [86]. IL-11 acts synergistically with IL-3 to stimulate CFU-GM formation and increases the number of progeny of CFU-GEMM both in vitro and in vivo [33,101,116]. We have found [84] that IL-11 induces the formation of osteoclasts in cultures of murine bone marrow and calvarial cells and stimulates bone resorption. In addition, we found that an antibody neutralizing IL-11 suppresses osteoclast development induced by either $1,25(OH)_2D_3$, PTH, IL-1 or TNF, whereas antibodies to IL-1 or TNF have no effect on the ability of IL-11 to induce osteoclast development. More important, in contrast to IL-6, the effects of IL-11 are independent of the estrogen status of the marrow donors, indicating that, as opposed to IL-6 which attains its importance for osteoclastogenesis only in pathologic states, IL-11 is involved in osteoclastogenesis under physiologic conditions.

M-CSF is another cytokine with a critical role for osteoclastogenesis, as evidenced by the fact that a deficiency in M-CSF is responsible for the paucity of osteoclasts in osteopetrotic op/op mice [123]. Lack of M-CSF in this condition is caused by an insertion mutation in the M-CSF gene, resulting in the formation of an incomplete coding region [126]. Administration of M-CSF to op/op mice increases osteoclast numbers, restores bone resorption, and cures the skeletal sclerosis [24]. M-CSF is also essential for both the proliferation and differentiation of osteoclast progenitors induced by $1,25(OH)_2D_3$ [112]; and in-

cubation of fetal bone with M-CSF causes increased osteoclast formation and bone resorption [17]. Osteoblastic cell lines and cultures of
osteoblastic cells from neonatal calvaria secrete M-CSF [20,40].

GM-CSF can induce the proliferation and maturation of hematopoietic progenitors such as CFU-GEMM, CFU-GM and their progeny
and stimulates osteoclast formation [66,105,110]. Nevertheless, GM-
CSF can also inhibit osteoclast formation [34,104,111]. This latter effect is probably due to a GM-CSF-mediated downregulation of receptors for M-CSF [120]. GM-CSF is secreted by osteoblastic cells upon
stimulation with PTH or lipopolysaccharide [38,39].

Finally, LIF, a multifunctional cytokine that influences certain aspects of hematopoietic cell differentiation [15], can also stimulate bone
resorption [46,93]. However, there is evidence that LIF can stimulate
bone formation as well. Mice injected with tumor cells expressing LIF
exhibit increased bone turnover and increased osteoblast formation
[73]. LIF also potentiates retinoic acid-induced expression of alkaline
phosphatase in preosteoblasts [96]. Osteoblasts produce LIF following
stimulation by IL-1 and TNF, but neither PTH nor $1,25(OH)_2D_3$ affect
LIF production by osteoblasts [45,46]. LIF is produced constitutively
by human bone marrow cells and can be further stimulated by IL-1 and
TGF-β [122].

6.4 Inhibition of IL-6 Production by Estrogens

Studies from this author's laboratory have indicated that 17β-estradiol
inhibits the production of bioassayable IL-6 (and the IL-6 mRNA) by
cultured bone marrow stromal and osteoblastic cell lines as well as primary bone cell cultures from rodents and humans [27]. This effect is
most likely mediated via the estrogen receptor (ER) [7]. Indeed, bone
marrow-derived stromal cells express ER at levels similar to those
found earlier in osteoblasts [22,58]. Moreover, nucleotide sequencing
of polymerase chain reaction (PCR)-amplified cDNA from bone marrow-derived stromal cells shows an essential identity between the bone
marrow stromal and the uterine ER. Consistent with a mediating role
of the ER in the regulation of IL-6 by 17β-estradiol, the pure estrogen
antagonist ICI 164,384 prevents the effect of 17β-estradiol on IL-6.
Further, in more recent studies of ours [91], we found that 17β-es-

tradiol completely suppresses stimulated transcription in HeLa cells cotransfected with constructs of the human IL-6 promoter linked to the reporter gene chloramphenicol acetyl transferase (CAT) and a human ER expression plasmid but has no effect on IL-6 transcription in HeLa cells not transfected with ER. Similarly, estradiol inhibits stimulated transcription from the human IL-6 promoter in transfected bone marrow stromal cells (which express the ER constitutively) without the requirement of cotransfection of the ER plasmid. These hormonal effects are indistinguishable between constructs containing a 1.2-kb fragment of the 5' flanking region of the human IL-6 gene promoter or only the proximal 225-bp segment. Nevertheless, yeast-derived recombinant ER does not bind nor does it compete with the binding of nuclear extracts to the 225-bp segment in DNA shift assays. These findings establish that estradiol inhibits IL-6 production by inhibiting the stimulated expression of the IL-6 gene through an ER-mediated indirect effect on the transcriptional activity of the proximal 225-bp sequence of the promoter, perhaps through an interference with events along the signaling pathways initiated by the IL-6 stimulating agents.

6.5 Upregulation of Osteoclastogenesis
Following Estrogen Loss and Its Mediation by IL-6

Based on the in vitro findings for the inhibition of IL-6 production by estrogen, we have tested in vivo the hypothesis that estrogen loss upregulates osteoclastogenesis through an increase in the production of IL-6 in the microenvironment of the marrow [49], using the ovariectomized mouse model [67]. We found that the number of CFU-GM colonies and the number of osteoclasts (identified by tartrate resistant acid phosphatase, TRAPase, staining and calcitonin binding, a combination of features unique to osteoclastic cells) was greater in short-term cultures of marrow cells from ovariectomized mice than in cultures from sham-operated animals [49]. An upregulating effect of ovariectomy in osteoclast-like (TRAPase positive cells) numbers formed in ex vivo cultures of marrow was also reported by Kalu [52]. The increased osteoclast formation in the marrow cultures in our studies was mirrored by an increase in the number of osteoclasts present in sections of trabecular bone. More important, administration of

17β-estradiol or injections of an IL-6 neutralizing antibody (but not administration of an immunoglobulin G isotype control antibody) to the ovariectomized animals prevented all these cellular changes. Consistent with these findings, we have found that estrogen loss causes an upregulation of IL-6 production by ex vivo bone marrow cell cultures in response to either $1,25(OH)_2D_3$ or PTH and that a similar phenomenon can be elicited in vitro by withdrawal of 17β-estradiol from primary cultures of calvarial cells [85].

In our published report [49], an increase in CFU-GM colonies following ovariectomy was also seen in cultures of spleen cells. More recent studies from our group [48] demonstrate that besides CFU-GM and osteoclasts, ovariectomy also causes an upregulation of other hematopoietic precursors known to be sensitive to IL-6, namely, CFU-GEMM and the burst-forming unit for erythrocytes (BFU-E). Similar to the case of CFU-GM, administration of either 17β-estradiol or a neutralizing antibody against IL-6 to the ovariectomized mice suppresses these changes. In line with the evidence for the regulation of IL-6 by estrogens, it has been found that an IL-6-dependent model of arthritis, namely, the pristane-induced arthritis, can be prevented during pregnancy [115]. In addition, there is evidence to indicate that estrogens suppress acquiring but not once acquired immunity [10] and that they have anti-inflammatory properties [51]. Hence, it is possible that the implications of cytokine regulation by estrogen might extend far beyond the upregulation of osteoclastogenesis.

6.6 Are Other Cytokines also Regulated by Estrogens?

Evidence from studies conducted primarily in peripheral blood mononuclear cells (PBMC) from humans raises the possibility that the production of IL-1, TNF, and GM-CSF might be regulated by estrogens, as it is the case for IL-6. Indeed, Pacifici et al. and Pioli et al. [78,81,89] have reported that PBMC from patients with high-turnover osteoporosis spontaneously produced IL-1, TNF, GM-CSF, and IL-6 and that PBMC from patients with high-turnover osteoporosis produce higher amounts of IL-1 than cells from low-turnover or nonosteoporotic patients. Treatment with ovarian steroids prevents the increased IL-1 release from monocytes from women with established osteoporosis [79,80]. In conflict with these findings, we have recently reported that unstimulated preparations of monocyte-enriched PBMC from osteoporotic women do not secrete IL-1β into the culture medium as measured by a specific enzyme-linked immunosorbent assay (ELISA) and that there is no difference in OKT3-stimulated IL-1β (or γ-interferon) production by PBMC between premenopausal and postmenopausal women (with and without osteoporosis) [42]. In addition, flow cytometric analysis of PBMC in our studies revealed no difference between osteoporotics and nonosteoporotics in the distribution of 18 phenotypic subsets, including CD4- or CD8-positive lymphocytes or the ratio of CD4 to CD8 cells, and no correlation between stimulated cytokine production or the surface markers with bone density. Similarly, Zarrabeitia et al. found no difference in IL-1β, TNF, or prostaglandin E_2 production by stimulated PBMC isolated from healthy postmenopausal women and patients with postmenopausal osteoporosis [127]. Whether the conflicting results regarding cytokine production by blood mononuclear cells from postmenopausal women are due to methodologic or other reasons currently remains unclear.

The apparent discrepancies regarding IL-1 production by PBMC notwithstanding, low levels of estradiol binding in rat peritoneal macrophages and in a human monocytic leukemia cell line have been reported by Gulshan et al. [31]. However, progesterone and estrogen have been shown to either induce or have no effect on IL-1 release from PBMC at physiologic concentrations, although progesterone, but not estradiol, did seem to inhibit the IL-1 mRNA [90]. Inhibition of unstimulated production of TNF by 17β-estradiol in monocytes from postmenopausal women has been reported by Ralston et al. [92]. This response, however, was not dose dependent and could not be observed in half of the postmenopausal women examined. Rickard et al. have reported a small and variable effect of 17β-estradiol on IL-1-stimulated TNF production by primary cultures of normal human osteoblast-like cells but no effect on IL-1 or IL-6 production; strangely, 17α-estradiol also had a small inhibitory effect on IL-1-induced TNF release [94]. Lack of an effect of 17β-estradiol on IL-6 production by primary cultures of normal human osteoblast-like cells has also been reported by Chaudhary et al. [16], raising the possibility that regulation of IL-6 by estrogen is not a universal response to all osteoblastic cells, but it might be limited to cells at certain stages of differentiation.

6.7 Concluding Remarks

To date, there is compelling evidence that the production of IL-6 by stromal and some, but not all, osteoblastic cells is inhibited by estrogens. Upon estrogen loss, this inhibitory effect is removed, resulting in an upregulation of IL-6 which, in turn, upregulates the formation of osteoclasts. The evidence that an anti-IL-6 antibody inhibits osteoclast activity in humans [57], taken together with the evidence for an inhibitory effect of 17β-estradiol on IL-6 at the protein level and the level of the human IL-6 gene promoter in human cells [27,91], establishes the relevance of the experimental evidence to humans. Based on this, I believe that the interrelationship between estrogen loss, upregulation of IL-6, and increased osteoclastogenesis provides a convincing mechanistic explanation for the pathophysiology of the bone loss that occurs in the postmenopausal state.

At this stage, it remains unclear whether IL-6 is the sole pathogenetic factor responsible for the increased osteoclastogenesis that ensues upon estrogen loss or whether other cytokines, such as IL-1 and TNF, are also upregulated and, therefore, contribute to this pathologic process. In my opinion, the evidence supporting the latter possibility is rather weak and circumstantial, especially since a mechanistic link between cytokine release by circulating blood mononuclear cells and the bone remodeling process is not apparent. In any event, it is reasonable to expect that the stimulatory signals for the increased IL-6 production in the estrogen-deficient state are provided by several agents, including PTH, $1,25(OH)_2D_3$, TGFβ, and IL-1 and TNF, since all of them are capable of stimulating IL-6 by stromal/osteoblastic cells. Considering this, removal of the inhibitory effect of estrogen on IL-6 is sufficient to account for the increased production of this cytokine and, therefore, the increased osteoclastogenesis, irrespective of whether IL-1 and TNF are also upregulated. Therefore, one need not invoke estrogenic regulation of IL-1 and TNF to explain the upregulation of osteoclastogenesis. The substantial quantitative difference in the production levels of IL-6 (nanomolar) versus IL-1 and TNF (picomolar) by stromal/osteoblastic cells and the evidence that the IL-1 and TNF signals for osteoclastogenesis are mediated via IL-6 and IL-11 (at least in part) renders a putative regulatory effect of estrogens on IL-1 and TNF even less critical. This, of course, does not exclude the possibility that IL-1 and

TNF are produced in the marrow in biologically significant quantities by the resident monocytes/macrophages, nor is it inconsistent with an important role of IL-1 and TNF in osteoclastogenesis, especially in inflammatory states such as the arthritides; however, it decreases the likelihood that IL-1 and TNF play a pathogenetic role in triggering the upregulation of osteoclastogenesis following loss of ovarian function.

Acknowledgements. The author wishes to acknowledge the support of the NIH (AR 41313) and the Department of Veterans Affairs for his research. He also wishes to acknowledge the contributions of his co-workers Drs. Robert L. Jilka, Giuseppe Girasole, Giovanni Passeri, Teresita Bellido, Hanna Mocharla, and Xiao-Peng Yu; his collaborators Hal Broxmeyer, John Abrams, Scott Pottratz, Scott Boswell, Daniel Williams, David Crabb, and Wesley Pike in the studies reviewed here; and the excellent secretarial support provided by Niki Barker in the writing of this article.

References

1. Agarwala N, Gay CV (1992) Specific binding of parathyroid hormone to living osteoclasts. J Bone Miner Res 7:531-539
2. Akatsu T, Takahashi N, Udagawa N, Sato K, Nagata N, Moseley JM, Martin TJ, Suda T (1989) Parathyroid hormone (PTH)-related protein is a potent stimulator of osteoclast-like multinucleated cell formation to the same extent as PTH in mouse marrow cultures. Endocrinology 125:20-27
3. Akira S, Hirano T, Taga T, Kishimoto T (1990) Biology of multifunctional cytokines: IL 6 and related molecules (IL 1 and TNF). FASEB J 4:2860-2867
4. Al-Humidan A, Ralston SH, Hughes DE, Chapman K, Aarden L, Russell RGG, Gowen M (1991) Interleukin-6 does not stimulate bone resorption in neonatal mouse calvariae. J Bone Miner Res 6:3-8
5. Ashton BA, Allen TD, Howlett CR, Eaglesom CC, Hattori A, Owen M (1980) Formation of bone and cartilage by marrow stromal cells in diffusion chambers in vivo. Clin Orthop Rel Res 151:294-307
6. Baron R, Vignery A (1981) Behavior of osteoclasts during a rapid change in their number induced by high doses of parathyroid hormone or calcitonin in intact rats. Bone 2:339-346
7. Bellido T, Girasole G, Passeri G, Yu XP, Mocharla H, Jilka RL, Notides A, Manolagas SC (1993) Demonstration of estrogen and vitamin D receptors in bone marrow derived stromal cells: upregulation of the estrogen receptor by 1,25(OH)$_2$D4$_3$. Endocrinology 133:553–562

8. Benayahu D, Kletter Y, Zipori D, Weintroub S (1989) Bone marrow-derived stromal cell line expressing osteoblastic phenotype in vitro and osteogenic capacity in vivo. J Cell Physiol 140:1-7

9. Benayahu D, Horowitz M, Zipori D, Wientroub S (1992) Hemopoietic functions of marrow-derived osteogenic cells. Calcif Tissue Int 51:195-201

10. Benten WPM, Wunderlich F, Mossmann H (1992) Plasmodium chabaudi: estradiol suppresses acquiring, but not once-acquired immunity. Exp Parasitol 75:240-247

11. Bertolini DR, Nedwin GE, Bringman TS, Smith DD, Mundy GR (1986) Stimulation of bone resorption and inhibition of bone formation in vitro by human tumour necrosis factors. Nature 319:516-518

12. Black K, Garrett IR, Mundy GR (1991) Chinese hamster ovarian cells transfected with the murine interleukin-6 gene cause hypercalcemia as well as cachexia, leukocytosis and thrombocytosis in tumor-bearing nude mice. Endocrinology 128:2657-2659

13. Boyce BF, Aufdemorte TB, Garrett IR, Yates AJ, Mundy GR (1989) Effects of interleukin-1 on bone turnover in normal mice. Endocrinology 125:1142-1150

14. Burger EH, Van der Meer JWM, Nijweide PJ (1984) Osteoclast formation from mononuclear phagocytes: role of bone-forming cells. J Cell Biol 99:1901-1906

15. Burstein SA, Mei RL, Henthorn J, Friese P, Turner K (1992) Leukemia inhibitory factor and interleukin-11 promote maturation of murine and human megakaryocytes in vitro. J Cell Physiol 153:305-312

16. Chaudhary LR, Spelsberg TC, Riggs BL (1992) Production of various cytokines by normal human osteoblast-like cells in response to interleukin-1β and tumor necrosis factor-α: lack of regulation by 17β-estradiol. Endocrinology 130:2528-2534

17. Corboz VA, Cecchini MG, Felix R, Fleisch H, van der Pluijm G, Löwik CWGM (1992) Effect of macrophage colony-stimulating factor on in vitro osteoclast generation and bone resorption. Endocrinology 130:437-442

18. Deldar A, Lewis H, Weiss L (1985) Bone lining cells and hematopoiesis: an electron microscopic study of canine bone marrow. Anat Rec 213:187-201

19. Dinarello CA (1991) Interleukin-1 and interleukin-1 antagonism. Blood 77:1627-1652

20. Elford PR, Felix R, Cecchini M, Trechsel U, Fleisch H (1987) Murine osteoblastlike cells and the osteogenic cell MC3T3-E1 release a macrophage colony-stimulating activity in culture. Calcif Tissue Int 41:151-156

21. Emerson SG (1991) The stem cell model of hematopoiesis. In: Hoffman R, Benz EJ, Shattil SJ, Furie B, Cohen HJ (eds) Hematology. Basic principles and practice. Churchill Livingstone, New York, pp 72-81
22. Eriksen EF, Colvard DS, Berg NJ, Graham ML, Mann KG, Spelsberg TC, Riggs BL (1988) Evidence of estrogen receptors in normal human osteoblast-like cells. Science 241:84-86
23. Eriksen EF, Hodgson SF, Eastell R, Cedel SL, O'Fallon WM, Riggs BL (1990) Cancellous bone remodeling in type I (postmenopausal) osteoporosis: quantitative assessment of rates of formation, resorption, and bone loss at tissue and cellular levels. J Bone Miner Res 5:311-319
24. Felix R, Cecchini MG, Fleisch H (1990) Macrophage colony stimulating factor restores in vivo bone resorption in the op/op osteopetrotic mouse. Endocrinology 127:2592-2594
25. Feyen JHM, Elford P, Dipadova RE, Trechsel U (1989) Interleukin-6 is produced by bone and modulated by parathyroid hormone. J Bone Miner Res 4:633-638
26. Friedenstein AJ, Chailakhjan RK, Latzinih NV, Panasyuk AF, Keiliss-Borok IV (1974) Stromal cells responsible for transferring the microenvironment of the hemopoietic tissues. Transplantation 17:331-339
27. Girasole G, Jilka RL, Passeri G, Boswell S, Boder G, Williams DC, Manolagas SC (1992) 17β-Estradiol inhibits interleukin-6 production by bone marrow-derived stromal cells and osteoblasts in-vitro: a potential mechanism for the antiosteoporotic effect of estrogens. J Clin Invest 89:883-891
28. Gowen M, Mundy GR (1986) Actions of recombinant interleukin 1, interleukin 2, and interferon-gamma on bone resorption in vitro. J Immunol 136:2478-2482
29. Gowen M, Chapman K, Littlewood A, Hughes D, Evans D, Russell G (1990) Production of tumor necrosis factor by human osteoblasts is modulated by other cytokines, but not by osteotropic hormones. Endocrinology 126:1250-1255
30. Graves L III, Jilka RL (1990) Comparison of bone and parathyroid hormone as stimulators of osteoclast development and activity in calvarial cell cultures from normal and osteopetrotic (mi/mi) mice. J Cell Physiol 145:102-109
31. Gulshan S, McCruden AB, Stimson WH (1990) Oestrogen receptors in macrophages. Scand J Immunol 31:691-697
32. Hagenaars CE, van der Kraan AA, Kawilarang-de Haas EW, Visser JW, Nijweide PJ (1989) Osteoclast formation from cloned pluripotent hemopoietic stem cells. Bone Miner 6:179-189

33. Hangoc G, Yin T, Cooper S, Schendel P, Yang Y-C, Broxmeyer HE (1993) In vivo effects of recombinant interleukin-11 on myelopoiesis in mice. Blood 81:965-972

34. Hattersley G, Chambers TJ (1990) Effects of interleukin 3 and of granulocyte-macrophage and macrophage colony stimulating factors on osteoclast differentiation from mouse hemopoietic tissue. J Cell Physiol 142:201-209

35. Hattersley G, Kerby JA, Chambers TJ (1991) Identification of osteoclast precursors in multilineage hemopoietic colonies. Endocrinology 128:259-262

36. Holtrop ME, Cox KA, Clark MR, Holick MF, Anast CS (1981) 1,25-Dihydroxy-cholecalciferol stimulates osteoclasts in rat bones in the absence of parathyroid hormone. Endocrinology 108:2293-2301

37. Horowitz MC, Jilka RL (1992) Colony stimulating factors and bone remodeling. In: Gowen M (ed) Cytokines and bone metabolism. CRC Press, Boca Raton, pp 185-227

38. Horowitz MC, Coleman DL, Flood PM, Kupper TS, Jilka RL (1989) Parathyroid hormone and lipopolysaccharide induce murine osteoblast-like cells to secrete a cytokine indistinguishable from granulocyte-macrophage colony-stimulating factor. J Clin Invest 83:149-157

39. Horowitz MC, Coleman DL, Ryaby JT, Einhorn TA (1989) Differential secretion of granulocyte-macrophage colony stimulating factor by a murine osteoblast cell line MC3T3. J Bone Miner Res 4:911-921

40. Horowitz MC, Einhorn TA, Philbrick W, Jilka RL (1989) Functional and molecular changes in colony stimulating factor secretion by osteoblasts. Connect Tissue Res 20:159-168

41. Horowitz M, Phillips J, Centrella M (1992) TGF-β regulates interleukin-6 secretion by osteoblasts. In: Cohn DV, Gennari C, Tashjian Jr AH (eds) Calcium regulating hormones and bone metabolism, vol 11. Excerpta Medica, Amsterdam, pp 275-280

42. Hustmyer FG, Walker E, Yu XP, Girasole G, Sakagami Y, Peacock M, Manolagas SC (1993) Cytokine production and surface antigen expression by peripheral blood mononuclear cells in postmenopausal osteoporosis. J Bone Miner Res 8:51-59

43. Ikebuchi K, Wong GG, Clark SC, Ihle JN, Hirai Y, Ogawa M (1987) Interleukin 6 enhancement of interleukin-3 dependent proliferation of multipotential hemopoietic progenitors. Proc Nat Acad Sci 84:9035-9039

44. Ishimi Y, Miyaura C, Jin CH, Akatsu T, Abe E, Nakamura Y, Yamaguchi A, Yoshiki S, Matsuda T, Hirano T, Kishimoto T, Suda T (1990) IL-6 is produced by osteoblasts and induces bone resorption. J Immunol 145:3297-3303

45. Ishimi Y, Abe E, Jin CH, Miyaura C, Hong MH, Oshida M, Kurosawa H, Yamaguchi Y, Tomida M, Hozumi M, Suda T (1992) Leukemia inhibitory factor/differentiation-stimulating factor (LIF/D-factor): regulation of its production and possible roles in bone metabolism. J Cell Physiol 152:71-78

46. Ishimi Y, Abe E, Jin CH, Miyaura C, Hong MH, Oshida M, Kurosawa H, Yamaguchi Y, Tomida M, Hozumi M (1992) Leukemia inhibitory factor/differentiation-stimulating factor (LIF/D-factor): regulation of its production and possible roles in bone metabolism. J Cell Physiol 152:71-78

47. Jilka RL, Passeri G, Girasole G, Marcus T, Manolagas SC (1991) Antibodies against tumor necrosis factor inhibit IL-1-induced IL-6 production in calvaria cells. J Bone Miner Res 6 (S1):S145

48. Jilka RL, Girasole G, Passeri G, Cooper S, Hangoc G, Abrams J, Broxmeyer H, Manolagas SC (1992) Ovariectomy in the mouse upregulates hematopoietic precursors in the bone marrow and their progeny in the peripheral blood: a mediating role of IL-6. J Bone Miner Res 7(S1):S115

49. Jilka RL, Hangoc G, Girasole G, Passeri G, Williams DC, Abrams JS, Boyce B, Broxmeyer H, Manolagas SC (1992) Increased osteoclast development after estrogen loss: mediation by interleukin-6. Science 257:88-91

50. Johnson RA, Boyce BF, Mundy GR, Roodman GD (1989) Tumors producing human tumor necrosis factor induced hypercalcemia and osteoclastic bone resorption in nude mice. Endocrinology 124:1424-1427

51. Josefsson E, Tarkowski A, Carlsten H (1992) Anti-inflammatory properties of estrogen. I. In vivo suppression of leukocyte production in bone marrow and redistribution of peripheral blood neutrophils. Cell Immunol 142:67-78

52. Kalu DN (1990) Proliferation of tartrate-resistant acid phosphatase positive multinucleate cells in ovariectomized animals. Proc Soc Exp Biol Med 195:70-74

53. Keeting PE, Rifas L, Harris SA, Colvard DS, Spelsberg TC, Peck WA, Riggs BL (1991) Evidence for interleukin-1β production by cultured normal human osteoblast-like cells. J Bone Miner Res 6:827-833

54. Keeting PE, Scott RE, Colvard DS, Anderson MA, Oursler MJ, Spelsberg TC, Riggs BL (1992) Development and characterization of a rapidly proliferating, well-differentiated cell line derived from normal adult human osteoblast-like cells transfected with SV40 large T antigen. J Bone Miner Res 7:127-136

55. Kerby JA, Hattersley G, Collins DA, Chambers TJ (1992) Derivation of osteoclasts from hematopoietic colony-forming cells in culture. J Bone Miner Res 7:353-361

56. King GJ, Holtrop ME, Raisz LG (1978) The relation of ultrastructural changes in osteoclasts to resorption in bone cultures stimulated with parathyroid hormone. Metab Bone Dis Rel Res 1:67-74

57. Klein B, Wijdenes J, Zhang XG, Jourdan M, Boiron JM, Brochier J, Liautard J, Merlin M, Clement C, Morel-Fournier B, Lu ZY, Mannoni P, Sany J, Bataille R (1991) Murine anti-interleukin-6 monoclonal antibody therapy for a patient with plasma cell leukemia. Blood 78:1198-1204

58. Komm BS, Terpening CM, Benz DJ, Graeme KA, Gallegos A, Korc M, Greene GL, O'Malley BW, Haussler MR (1988) Estrogen binding, receptor mRNA, and biologic response in osteoblast-like osteosarcoma cells. Science 241:81-84

59. Kurihara N, Chenu C, Miller M, Civin C, Roodman GD (1990) Identification of committed mononuclear precursors for osteoclast-like cells formed in long term human marrow cultures. Endocrinology 126:2733-2741

60. Kurihara N, Civin C, Roodman GD (1991) Osteotropic factor responsiveness of highly purified populations of early and late precursors for human multinucleated cells expressing the osteoclast phenotype. J Bone Miner Res 6:257-261

61. Li Y, Matsuzaki N, Masuhiro K, Kameda T, Taniguchi T, Saji F, Yone K, Tanizawa O (1992) Trophoblast-derived tumor necrosis factor-alpha induces release of human chorionic gonadotropin using interleukin-6 (IL-6) and IL-6-receptor-dependent system in the normal human trophoblasts. J Clin Endocrinol Metab 74:184-191

62. Linkhart TA, Linkhart SG, MacCharles DC, Long DL, Strong DD (1991) Interleukin-6 messenger RNA expression and interleukin-6 protein secretion in cells isolated from normal human bone: regulation by interleukin-1. J Bone Miner Res 6:1285-1294

63. Littlewood AJ, Russell J, Harvey GR, Hughes DE, Russell RGG, Gowen M (1991) The modulation of the expression of IL-6 and its receptor in human osteoblasts in vitro. Endocrinology 129:1513-1520

64. Liu CC, Howard GA (1991) Bone-cell changes in estrogen-induced bone-mass increase in mice: dissociation of osteoclasts from bone surfaces. Anat Rec 229:240-250

65. Löwik CWGM, van der Pluijm G, Bloys H, Hoekman K, Bijvoet OLM, Aarden LA, Papapoulos SE (1989) Parathyroid hormone (PTH) and PTH-like protein (PLP) stimulate interleukin-6 production by osteogenic cells: a possible role of interleukin-6 in osteoclastogenesis. Biochem Biophys Res Comm 162:1546-1552

66. MacDonald BR, Mundy GR, Clark S, Wang EA, Keuhl TJ, Stanley ER, Roodman GD (1986) Effects of human recombinant CSF-GM and highly purified CSF-1 on the formation of multinucleated cells with osteoclast

characteristics in long-term bone marrow cultures. J Bone Miner Res 1:227-233

67. Manolagas SC, Jilka RL (1992) Cytokines, hematopoiesis, osteoclastogenesis, and estrogens. Calcif Tissue Int 50:199-202

68. Marks Jr SC, Walker DG (1976) Mammalian osteopetrosis – a model for studying cellular and humoral factors in bone resorption. In: Bourne GH (ed) The biochemistry and physiology of bone, vol IV. Academic Press, New York, pp 227-301

69. McCulloch CA, Strugurescu M, Hughes F, Melcher AH, Aubin JE (1991) Osteogenic progenitor cells in rat bone marrow stromal populations exhibit self-renewal in culture. Blood 77:1906-1911

70. McSheehy PM, Chambers TJ (1986) Osteoblastic cells mediate osteoclastic responsiveness to parathyroid hormone. Endocrinology 118:824-828

71. McSheehy PM, Chambers TJ (1987) 1,25-Dihydroxyvitamin D3 stimulates rat osteoblastic cells to release a soluble factor that increases osteoclastic bone resorption. J Clin Invest 80:425-429

72. Metcalf D (1989) The molecular control of cell division, differentiation commitment and maturation in haemopoietic cells. Nature 339:27-30

73. Metcalf D, Gearing DP (1989) Fatal syndrome in mice engrafted with cells producing high levels of the leukemia inhibitory factor. Proc Natl Acad Sci USA 86:5948-5952

74. Miller SC, Bowman BM, Myers RL (1984) Morphological and ultrastructural aspects of the activation of avian medullary bone osteoclasts by parathyroid hormone. Anat Rec 208:223-231

75. Neta R, Perlstein R, Vogel SN, Ledney GD, Abrams J (1992) Role of interleukin 6 (IL-6) in protection from lethal irradiation and in endocrine responses to IL-1 and tumor necrosis factor. J Exp Med 175:689-694

76. Ohsaki Y, Takahashi S, Scarcez T, Demulder A, Nishihara T, Williams R, Roodman GD (1992) Evidence for an autocrine/paracrine role for interleukin-6 in bone resorption by giant cells from giant cell tumors of bone. Endocrinology 131:2229-2234

77. Owen M (1985) Lineage of osteogenic cells and their relationship to the stromal system. In: Peck WA (ed) Bone and mineral research, vol. 3. Elsevier, Amsterdam, pp1-25

78. Pacifici R, Rifas L, Teitelbaum S, Slatopolsky E, McCracken R, Bergfeld M, Lee W, Avioli LV, Peck WA (1987) Spontaneous release of interleukin 1 from human blood monocytes reflects bone formation in idiopathic osteoporosis. Proc Natl Acad Sci USA 84:4616-4620

79. Pacifici R, Rifas L, McCracken R, Vered I, McMurtry C, Avioli LV, Peck WA (1989) Ovarian steroid treatment blocks a postmenopausal increase in blood monocyte interleukin 1 release. Proc Natl Acad Sci USA 86:2398-2402

80. Pacifici R, Brown C, Puscheck E, Friedrich E, Slatopolsky E, Maggio D, McCracken R, Avioli LV (1991) Effect of surgical menopause and estrogen replacement on cytokine release from human blood mononuclear cells. Proc Natl Acad Sci USA 88:5134-5138

81. Pacifici R, Brown C, Puscheck E, Friedrich E, Slatopolsky E, Maggio D, McCracken R, Avioli LV (1991) Effect of surgical menopause and estrogen replacement on cytokine release from human blood mononuclear cells. Proc Natl Acad Sci USA 88:5134-5138

82. Parfitt AM (1983) The physiologic and clinical significance of bone histomorphometric data. In: Recker RR (ed) Bone histomorphometry: techniques and interpretation. CRC Press, Boca Raton, pp 143-223

83. Parfitt AM, Mathews CHE, Villanueva AR, Kleerekoper M, Frame B, Rao DS (1983) Relationship between surface, volume and thickness of iliac trabecular bone in aging and in osteoporosis: implications for the microanatomic and cellular mechanism of bone loss. J Clin Invest 72:1396-1409

84. Passeri G, Girasole G, Knutson S, Yang YC, Manolagas SC, Jilka RL (1992) Interleukin-11 (IL-11): a new cytokine with osteoclastogenic and bone resorptive properties and a critical role in PTH- and $1,25(OH)_2D_3$-induced osteoclast development. J Bone Miner Res 7(S1):S110(Abstract)

85. Passeri G, Girasole G, Jilka RL, Manolagas SC (1993) Increased IL-6 production by murine bone marrow and bone cells following estrogen withdrawal. Endocrinology 133:822–828

86. Passeri G, Bellido T, Girasole G, Tkaczyk A, Manolagas SC, Jilka RL (1993) Transforming growth factor-β (TGFβ) and interleukin-1 (IL-1) induce the interleukin-11 (IL-11) mRNA in both bone marrow-derived stromal cells and osteoblasts from humans. J Bone Miner Res 8(S1) (in press) (Abstract)

87. Paul SR, Bennett F, Calvetti JA, Kelleher K, Wood CR, O'Hara RM Jr, Leary AC, Sibley B, Clark SC, Williams DA, Yang YC (1990) Molecular cloning of a cDNA encoding interleukin 11, a stromal cell-derived lymphopoietic and hematopoietic cytokine. Proc Natl Acad Sci USA 87:7512-7516

88. Pfeilschifter J, Chenu C, Bird A, Mundy GR, Roodman GD (1989) Interleukin-1 and tumor necrosis factor stimulate the formation of human osteoclastlike cells in vitro. J Bone Miner Res 4:113-118

89. Pioli G, Basini G, Pedrazzoni M, Musetti G, Ulietti V, Bresciani D, Villa P, Bacchi A, Hughes D, Russell G, Passeri M (1992) Spontaneous release of interleukin-1 and interleukin-6 by peripheral blood mononuclear cells after oophorectomy. Clin Sci 83:503-507

90. Polan ML, Danieli A, Kuo A (1988) Gonadal steroids modulate human monocyte interleukin-1 (IL-1) activity. Fertil Steril 49:964-968

91. Pottratz S, Bellido T, Mocharla H, Girasole G, Jilka R, Manolagas S, Crabb D (1992) 17β-Estradiol inhibits stimulated transcription from the human IL-6 promoter in transfected HeLa and murine bone marrow stromal cells. J Bone Miner Res 7(S1):S126

92. Ralston SH, Russell RGG, Gowen M (1990) Estrogen inhibits release of tumor necrosis factor from peripheral blood mononuclear cells in postmenopausal women. J Bone Miner Res 5:983-988

93. Reid LR, Lowe C, Cornish J, Skinner SJ, Hilton DJ, Willson TA, Gearing DP, Martin TJ (1990) Leukemia inhibitory factor: a novel bone-active cytokine. Endocrinology 126:1416-1420

94. Rickard D, Russell G, Gowen M (1992) Oestradiol inhibits the release of tumor necrosis factor but not interleukin 6 from adult human osteoblasts in vitro. Osteoporos Int 2:94-102

95. Rodan SB, Wesolowski G, Chin J, Limjuco GA, Schmidt JA, Rodan GA (1990) IL-1 binds to high affinity receptors on human osteosarcoma cells and potentiates prostaglandin E_2 stimulation of cAMP production. J Immunol 145:1231-1237

96. Rodan SB, Wesolowski G, Hilton DJ, Nicola NA, Rodan GA (1990) Leukemia inhibitory factor binds with high affinity to preosteoblastic RCT-1 cells and potentiates the retinoic acid induction of alkaline phosphatase. Endocrinology 127:1602-1608

97. Roodman GD, Kurihara N, Ohsaki Y, Kukita A, Hosking D, Demulder A, Smith JF, Singer FR (1992) Interleukin 6. A potential autocrine/paracrine factor in Paget's disease of bone. J Clin Invest 89:46-52

98. Rouleau MF, Warshawsky H, Goltzman D (1986) Parathyroid hormone binding in vivo to renal, hepatic, and skeletal tissues of the rat using a radioautographic approach. Endocrinology 118:919-931

99. Rouleau MF, Mitchell J, Goltzman D (1988) In vivo distribution of parathyroid hormone receptors in bone: evidence that a predominant osseous target cell is not the mature osteoblast. Endocrinology 123:187-192

100. Sakagami Y, Girasole G, Yu XP, Boswell HS, Manolagas SC (1993) Stimulation of interleukin-6 production by either calcitonin gene-related peptide or parathyroid hormone in two phenotypically distinct bone marrow-derived murine stromal cell lines. J Bone Miner Res 8:811–816

101. Schibler KR, Yang YC, Christensen RD (1992) Effect of interleukin-11 on cycling status and clonogenic maturation of fetal and adult hematopoietic progenitors. Blood 80:900-903

102. Serke S, Säuberlich S, Abe Y, Huhn D (1991) Analysis of CD34-positive hemopoietic progenitor cells from normal human adult peripheral blood: flow-cytometrical studies and in-vitro colony (CFU-GM, BFU-E) assays. Ann Hematol 62:45-53

103. Shadduck RK, Waheed A, Greenberger JS, Dexter TM (1983) Production of colony stimulating factor in long-term bone marrow cultures. J Cell Physiol 114:88-92
104. Shinar DM, Sato M, Rodan GA (1990) The effect of hemopoietic growth factors on the generation of osteoclast-like cells in mouse bone marrow cultures. Endocrinology 126:1728-1735
105. Sieff CA, Emerson SG, Donahue RE, Nathan DG, Wang EA, Wong GG, Clark SC (1985) Human recombinant granulocyte-macrophage colony-stimulating factor: a multilineage hematopoietin. Science 230:1171-1173
106. Starnes HF Jr, Pearce MK, Tewari A, Yim JH, Zou JC, Abrams JS (1990) Anti-IL-6 monoclonal antibodies protect against lethal Escherichia coli infection and lethal tumor necrosis factor-α challenge in mice. J Immunol 145:4185-4191
107. Stashenko P, Dewhirst FE, Peros WJ, Kent RL, Ago JM (1987) Synergistic interactions between interleukin 1, tumor necrosis factor, and lymphotoxin in bone resorption. J Immunol 138:1464-1468
108. Suda T, Miyaura C, Abe E, Kuroki T (1986) Modulation of cell differentiation, immune responses and tumor promotion by vitamin D compounds. In: Peck WA (ed) Bone and mineral research, vol 4. Elsevier, Amsterdam, pp 1-48
109. Suda T, Takahashi N, Martin TJ (1992) Modulation of osteoclast differentiation. Endocr Rev 13:66-80
110. Takahashi N, Yamana H, Yoshiki S, Roodman GD, Mundy GR, Jones SJ, Boyde A, Suda T (1988) Osteoclast-like cell formation and its regulation by osteotropic hormones in mouse bone marrow cultures. Endocrinology 122:1373-1382
111. Takahashi N, Udagawa N, Akatsu T, Tanaka H, Shionome M, Suda T (1991) Role of colony-stimulating factors in osteoclast development. J Bone Miner Res 6:977-985
112. Tanaka S, Takahashi N, Udagawa N, Tamura T, Akatsu T, Stanley ER, Kurokawa T, Suda T (1993) Macrophage colony-stimulating factor is indispensable for both proliferation and differentiation of osteoclast progenitors. J Clin Invest 91:257-263
113. Thomson BM, Saklatvala J, Chambers TJ (1986) Osteoblasts mediate interleukin 1 stimulation of bone resorption by rat osteoclasts. J Exp Med 164:104-112
114. Thomson BM, Mundy GR, Chambers TJ (1987) Tumor necrosis factors alpha and beta induce osteoblastic cells to stimulate osteoclastic bone resorption. J Immunol 138:775-779
115. Thompson SJ, Hitsumoto Y, Zhang YW, Rook GAW, Elson CJ (1992) Agalactosyl IgG in pristane-induced arthritis. Pregnancy affects the incidence and severity of arthritis and the glysocylation status of IgG. Clin Exp Immunol 89:434-438

116. Tsuji K, Lyman SD, Sudo T, Clark SC, Ogawa M (1992) Enhancement of murine hematopoiesis by synergistic interactions between steel factor (ligand for c-kit), interleukin-11, and other early acting factors in culture. Blood 79:2855-2860

117. Turner RT, Wakley GK, Hannon KS, Bell NH (1988) Tamoxifen inhibits osteoclast-mediated resorption of trabecular bone in ovarian hormone-deficient rats. Endocrinology 122:1146-1150

118. Udagawa N, Takahashi N, Akatsu T, Tanaka H, Sasaki T, Nishihara T, Koga T, Martin TJ, Suda T (1990) Origin of osteoclasts: mature monocytes and macrophages are capable of differentiating into osteoclasts under a suitable microenvironment prepared by bone marrow-derived stromal cells. Proc Natl Acad Sci USA 87:7260-7264

119. Vassalli P (1992) The pathophysiology of tumor necrosis factors. Annu Rev Immunol 10:411-452

120. Walker F, Nicola NA, Metcalf D, Burgess AW (1985) Hierarchical down regulation of hemopoietic growth factor receptors. Cell 43:269-276

121. Westen H, Bainton DF (1979) Association of alkaline-phosphatase-positive reticulum cell in bone marrow with granulocyte precursors. J Exp Med 50:919-937

122. Wetzler M, Talpaz M, Lowe DG, Baiocchi G, Gutterman JU, Kurzrock R (1991) Constitutive expression of leukemia inhibitory factor RNA by human bone marrow stromal cells and modulation by IL-1, TNF-α, and TGF-β. Exp Hematol 19:347-351

123. Wiktor-Jedrzejczak W, Bartocci A, Ferrante AW Jr, Ahmed-Ansari A, Sell KW, Pollard JW, Stanley ER (1990) Total absence of colony-stimulating factor 1 in the macrophage-deficient osteopetrotic (op/op) mouse. Proc Natl Acad Sci USA 87:4828-4832

124. Wong GG, Witek JAS, Temple PA, Kriz R, Ferenz C, Hewick RM, Clark SC, Ikebuchi K, Ogawa M (1988) Stimulation of murine hematopoietic colony formation by human interleukin-6. J Immunol 140:3040-3044

125. Yamashita T, Asano K, Takahashi N, Akatsu T, Udagawa N, Sasaki T, Martin TJ, Suda T (1990) Cloning of an osteoblastic cell line involved in the formation of osteoclast-like cells. J Cell Physiol 145:587-595

126. Yoshida H, Hayashi SI, Kunisada T, Ogawa M, Nishikawa S, Okamura H, Sudo T, Shultz LD, Nishikawa SI (1990) The murine mutation osteopetrosis is in the coding region of the macrophage colony stimulating factor gene. Nature 345:442-444

127. Zarrabeitia MT, Riancho JA, Amado JA, Napal J, Gonzalez-Macias J (1991) Cytokine production by peripheral blood cells in postmenopausal osteoporosis. Bone Miner 14:161-167

7 Sex Steroids and Prostaglandins in Bone Metabolism

Webster S.S. Jee, Yanfei F. Ma, Mei Li, Xiaoquang G. Liang,
Baiyun Y. Lin, Xiaojian J. Li, Hua Z. Ke, Satoshi Mori,
Rebecca B. Setterberg, and Donald B. Kimmel

7.1 Introduction

Sex steroids play an important role in skeletal growth and maintenance of bone mass in adults (Johnston 1985). Estrogen deficiency is important in the development of osteoporosis in women. Androgen deficiency may be as important in osteoporotic men (Foresta et al. 1985). The main effects of estrogen and androgen withdrawal are an increase in bone resorption with a smaller increase in bone formation and a consequent decrease in bone mass (osteopenia or osteoporosis; Jackson et al. 1987).

Raising bone mass and improving bone architecture of osteoporotic patients seems the best way to cure osteoporosis. Saving the remaining bone by reducing its turnover rate seems a poor second choice. One possible approach for reversing existing osteopenia is to use a bone anabolic agent to restore lost bone. This is necessary because osteoporotic patients always have either low bone mass or poor bone structure and increased fracture risk at numerous sites. It has been shown in human, dog and rat skeletons that intermittent administration of a bone anabolic agent such as prostaglandin E_2 (PGE_2) stimulates positive bone balance (Norrdin et al. 1990). We have reported many studies using PGE_2 in intact and ovariectomized female rats. Less is known of the effects of PGE_2 in male and orchiectomized rats, especially at the tissue and cellular level by histomorphometry.

This report summarizes our experiments dealing with the role of the bone anabolic agent PGE_2 in intact male and female rats and in preventing and curing estrogen deficiency and preventing immobilization-induced osteopenia in the rat. It will include findings of studies in progress dealing with the impact of PGE_2 upon intact male and female and ovariectomized and orchiectomized rats.

7.2 The Approach

7.2.1 The Dog and Rat Skeleton

The use of the rat for preclinical studies has obvious advantages in cost, ready availability, extent of background information on origin and care, and the rapidity of bone response (Jee 1991). Less apparent is that the mechanisms which control gains in bone mass (longitudinal bone growth and modeling drifts) and losses in bone mass (bone-metabolic, unit-based remodeling) are the same in young and old rats and humans. Furthermore, the rat and human skeletons respond similarly to mechanical loading, hormones, drugs, and other agents (Frost and Jee 1992).

The other skeleton employed is the adult male Beagle dog skeleton, an excellent experimental analog of the adult human skeleton – in particular as an experimental model for studies requiring Haversian and cancellous bone remodeling when estrogen is not a factor (Kimmel 1991). The role of the Beagle dog skeleton in studying estrogen depletion-induced bone loss is limited because of the small skeletal effect of oophorectomy (Kimmel 1991). Besides, doing a proper study with dogs is costly.

7.2.2 Bone Histomorphometry

Past studies of the intact male and female skeletons and the effects of sex steroids on bone were limited to measurements of ash and dry weights, calcium content, specific gravity, and external densitometry. These methods provide information about the net changes in bone mass and cannot provide information on the site-specific tissue and cellular changes of the intermediary organization processes involving growth, modeling, and remodeling (Frost 1986; Jee et al. 1983). Histomorphometry bridges the gap between cell and molecular biology and clinical investigations. Parfitt (1990) states that "although dynamic histomorphometry is unable to identify the primary target of any skeletally active agent, for determining the cumulative summation of all effects, whether primary or secondary, direct or indirect, immediate or remote, it is not merely a good way; it is, in the present state of knowledge, the only way."

7.2.3 The University of Utah Protocol

The majority of our studies have employed the rat skeleton between 7 weeks and 15 months of age. The objective of these studies were to test whether PGE_2 can prevent or cure ovariectomy- and immobilization-induced osteopenia. Various sites were studied: proximal tibial metaphysis (PTM), distal tibial metaphysis (DTM), fourth lumbar vertebral body (LVB) and the tibial shaft (TX) proximal to the tibiofibular junction. At 3 months of age, PTM and LVB were slow-growing bone sites, the former faster than the latter (Li et al. 1991). Although their metaphyses cannot be considered as adult cancellous bone sites, generally the secondary spongiosa can be studied if the longitudinal growth rate is low enough so as not to add new bone to the secondary spongiosa. An added bonus is that these two sites allow one to study the effect of an agent on longitudinal bone growth (Thorngren and Hansson 1973). The DTM is an adult bone site, its epiphysis closes at 3 months (Dawson 1925; Ito et al. 1993). Lastly, the TX proximal to the tibiofibular junction is a cortical bone site which allows the study of the formation drift (activation to formation, A→F; direct bone formation), the rearrangement of cortical bone by radial bone growth, the enlargement of the marrow cavity, and intracortical remodeling. Intracortical remodeling does not normally occur in the rat unless there are strong bone resorption stimuli (Ruth 1953; de Winter and Steendijk 1975). Another unique thing about this site is that the periosteal surface apposes bone by formation drift and the endocortical surface by modeling and remodeling. We routinely performed static and dynamic histomorphometry and microanatomical structural analyses on double fluorescent labeled undecalcified sections (Frost 1977; Jee et al. 1983; Parfitt et al. 1987; Compston et al. 1987; Thorngren and Hansson 1973). The transient and new steady state responses were studied in the same or separate protocol (Frost 1986; Li et al. 1990a).

The male adult Beagle skeletons were employed in two studies: (1) The effects of graded doses of PGE_2 by ^{99m}Tc diphosphonate scan (Fogelman et al. 1978) and (2) the transient effects of PGE_2 on iliac crest biopsy (cancellous bone) and the tibial shaft (Li et al. 1990b).

7.2.4 The Rat Osteopenia Models

We employed the standard ovariectomy and orchiectomy models to induce cancellous bone osteopenia (Waynforth 1980; Kalu 1991; Wronski and Yen 1991; Wakley and Turner 1991). We also used the right hindlimb immobilization (RHLI) model to induce a cancellous bone osteopenia. Since many are not familiar with this model, a brief description follows: The rats' hindlimbs were immobilized using a variation of Lindgren's method (1976). The right hindlimb was underloaded by immobilizing it against the abdomen using four layers of elastic bandage tape (Elastikon, Johnson and Johnson, New Brunswick, NJ) with the hip joint in flexion and the knee and ankle joints in extension so that the right hindlimb was immobilized and the left hindlimb was overloaded by the weight normally shared by two hindlimbs during ambulation. Within a day of their immobilization, all of the rats adapted to their new condition or walked or hopped about on three limbs (Jee et al. 1991b).

7.3 Effect of PGE_2 In Vivo

Earlier studies convinced us that PGE_2 is a suitable bone-forming agent for preclinical testing of agents to prevent and cure osteoporosis. Though the vigorous in vivo hard tissue anabolic effects of PGE_2 seem different from some of its in vitro effects, the anabolic effect data from dogs, humans, and rats are compelling (Norrdin et al. 1990). Woven bone formation exists periosteally in human infants given PGE_1 and dogs and rats given PGE_2 (Norrdin et al. 1990).

It not only stimulates formation of thicker trabeculae and new woven bone trabeculae in the marrow cavities of tibial metaphyses, but also adds new lamellar and woven bone to endocortical and periosteal surfaces in the tibial shaft (Ueno et al. 1985; Jee et al. 1985, 1987).

7.4 Inventory and Summary of University of Utah Findings

The summary of our findings is followed by a brief description of the following topics: (1) the transient and new steady state from continuous daily treatment; (2) the consequences of withdrawal of PGE_2 treatment, (3) the distribution of PGE_2-induced changes in the rat and the adult Beagle dog skeleton; (4) the lose, restore and maintain (LRM) and the supplement and maintain (SM) concepts to cure established osteopenia; (5) the skeletal action of PGE_2 upon the intact and castrated male and female skeletons, and (6) the contributions of our studies to a better understanding of bone biology.

7.4.1 Summary of Findings

1. Effects of extra PGE$_2$ in vivo
    - Activates woven bone formation, the highest level of osteoblastic activity
    - Stimulates direct bone formation during modeling and remodeling
    - Stimulates modeling and remodeling-dependent bone gain in a dose-effect relationship
    - Activates bone remodeling with net bone gain (formation resorption)
    - Activates intracortical remodeling normally not seen in rat bone
    - Stimulates periosteal bone formation to increase mass and strength
    - Stimulates endosteal bone formation (trabecular and endo-cortical) to manufacture more and thicker trabeculae, increase cortical diameter and decrease the size of the marrow cavity
    - Stimulates woven and lamellar marrow trabecular bone deposition to decrease the size of the marrow cavity
    - Shortens all phases of the remodeling cycle
    - Stimulates longitudinal bone growth in female rats, but depresses it in male rats
    - Depresses body and muscle weights and increases adrenal weights in male and female rats (role of adrenal androgens?)
    - Increases bone mass in the following order: tibial shaft > distal tibial metaphysis (adult) > proximal tibial metaphysis (slow growing) >> lumbar vertebral body (slower growing)
    - Stimulates woven and lamellar bone formation in regions of the skeleton dominated by cortical bone and low turnover cancellous bone more than in regions dominated by high turnover cancellous bone
    - Stimulates bone formation even when given with a strong anti-resorptive agent, a bisphosphonate
    - Continuous administration establishes a new steady state of increased bone mass with increased turnover rate
    - Withdrawal returns the previous steady state of bone mass and lower turnover rate
    - Adds bone in rats and dogs of all ages

2. Effects of PGE_2 on experimental osteopenic models
   - Prevents cancellous bone loss in ovariectomized rats
   - Adds extra cortical bone in ovariectomized rats.
   - Restores cancellous bone in rats with established ovariectomy-induced cancellous bone osteopenia
   - Adds extra cortical bone in rats with established ovariectomy-induced cancellous bone osteopenia
   - Stimulates longitudinal bone growth in ovariectomized rats
   - Depresses muscle weights in ovariectomized rats
   - Stimulates adrenal weights in ovariectomized rats
   - Prevents cancellous and cortical bone loss in hindlimb immobilized rats
   - Adds extra cancellous bone in hindlimb immobilized rats
   - Restores cancellous and cortical bone to bone loss by immobilization
   - Depresses muscle weights in immobilized rats
   - Increases adrenal weights in immobilized rats.

3. Effects of the cyclical use of PGE_2 and a bisphosphonate on curing osteopenia
   - Reverses and maintains cancellous bone osteopenia using the LRM concept
   - Increases cortical bone mass and bone architecture using the SM concept

4. Effects of PGE_2 on male and female skeletons
   - Adds more bone in male than the female rats
   - Depresses longitudinal bone growth in male rats
   - Further depresses longitudinal bone growth in orchiectomized rats
   - Stimulates bone formation equally in castrated rats

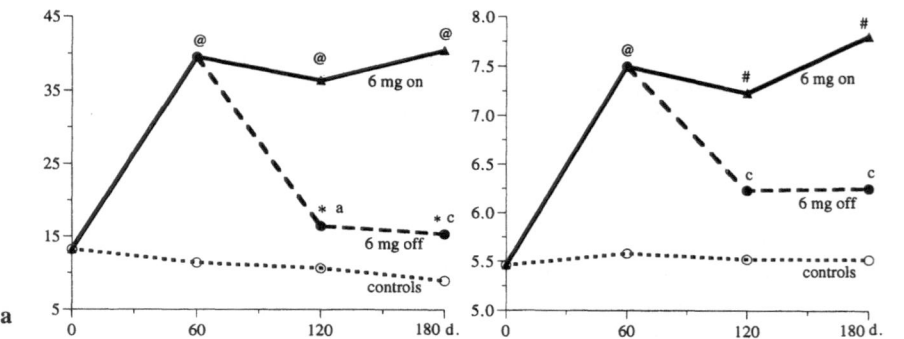

Fig. 1a–d. PGE$_2$-induced increase in bone that disappears after treatment is stopped (6 mg off) in the proximal tibial metaphysis **(a,c)** and tibial shaft **(b,d)**. Microradiographs showing cancellous bone changes in proximal tibial metaphyses **(c)** and tibial shaft **(d)** from age-matched controls at day 60 *(A)*, 120 *(B)*, and 180 *(C)*, and from rats treated with PGE$_2$ at 6 mg/kg per day for 60 days *(D and G)* 120 *(E)*, and 180*(F)* days, 60 days on(treated) + 60> days off (not treated; *H)*, and 60 days on and 120 days off *(I)*. *$p < 0.05$ and @$p < 0.01$ vs. controls; $^a p < 0.05$ and $^c p < 0.001$ vs. 60-day PGE$_2$-treated. × 4.5. (From Ke et al. 1991, 1992a; Jee et al. 1991a, 1992a)

7.4.2 Skeletal Response to Extra PGE$_2$ In Vivo

7.4.2.1 ˙Transient and New Steady State from Continuous Daily Treatment

PGE$_2$ stimulated direct bone formation at all surfaces, increased activation frequency to increase remodeling in the formation mode (i.e., bone formation exceeded bone resorption), deposited new woven, laminar and lamellar bone, triggered intracortical bone remodeling and shortened the bone remodeling period (Ueno et al. 1985; Jee et al. 1985, 1987, 1990, 1991a,b; Mori et al. 1990,1992; Li et al. 1990b, 1993; Ke et al. 1992a,b; Akamine et al. 1992; Tang et al. 1992; Ito et al. 1993; Tables 1,2; Figs. 1,2; see also Fig. 6). Tables 1 and 2 list the histomorphometry parameters in support of the above findings. Intracortical remodeling is uncommon in rats unless it is provoked by low calcium diet along with lactation (Ruth 1953; de Winter and Steendijk 1975). Thus when it does happen, then PGE$_2$ can be said to unequivocally stimulate bone resorption. Continual PGE$_2$ administration is as-

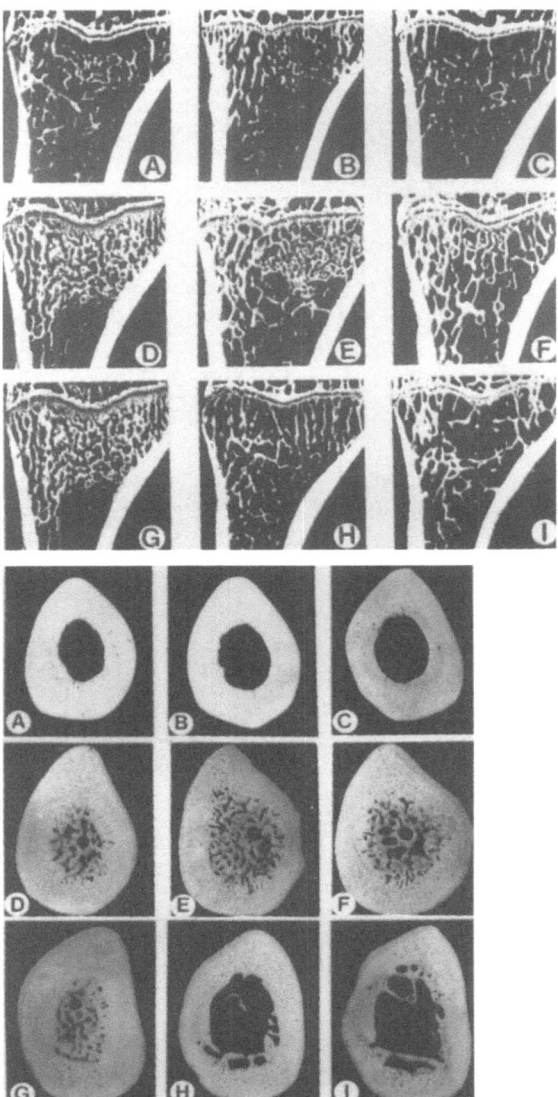

c

d

Table 1. Changes in the secondary spongiosa of proximal tibial metaphyseal histomorphometry of PGE$_2$-treated male rats[a]

Group	Trabecular area (%)	Trabecular width (mm)	Trabecular number (n/mm)	Eroded perimeter (%)	Bone formation rate/BV (%/year)	Remodeling period (day)
Basal	13.21	57.24	2.30	6.26	330	42.27
SD	2.68	6.98	0.34	0.85	147	3.68
60 days, aging	11.36	51.46	2.20	7.09	404	42.36
SD	4.75	9.01	0.84	1.72	191	7.50
%-B	−14	−10	−4	13	22	0
60 days PGE$_2$	39.49	78.73	5.02	9.13	722	31.86
SD	7.18	8.71	0.75	0.39	138	3.24
%-B	199**[b]	38**	118**	46**	119**	−25**
%-60A	248**	53**	128**	29**	79**	−25*

%B = compared to basal; %60A = compared to 60-day aging rats

*$p < 0.05$; **$p < 0.01$

[a]Seven-month-old male rats treated with 6 mg PGE$_2$/kg per day for 60 days (Ke et al. 1992a)

[b]Compared to basal controls (%-B) or age-related controls (%60A).

Table 2. Changes in tibial diaphyseal histomorphometry in PGE$_2$-treated male rats[a]

Group	Total tissue area (mm^2)	Marrow trabecular area (mm^2)	Intracortical porosity area (mm^2)	Total bone area (mm^2)	Marrow space area (mm^2)	Periosteal bone formation rate (μm/day* $\times$ 100)	Endocortical bone formation rate (μm/day* $\times$ 100)
Basal	6.62	0.000	0.000	5.46	1.16	21.30	21.10
SD	0.67	0.000	0.000	0.67	0.16	7.43	2.54
60 day, aging	6.91	0.000	0.000	5.58	1.33	22.19	23.23
SD	0.66	0.000	0.000	0.71	0.05	6.77	10.69
%-B	4				158	4	10
60 day PGE$_2$	8.10	1.103	0.142	7.64	0.46	91.54	80.59
SD	0.86	0.258	0.128	0.87	0.26	23.62	15.55
%60A	17*	**	**	37*	-67*	313**	247**

%B = compared to basal; %60A = compared to 60-day aging rats

* p < 0.05; ** p < 0.01

[a] Seven-month-old male rats treated with 6 mg PGE$_2$/kg per day for 60 days (Jee et al. 1992a)

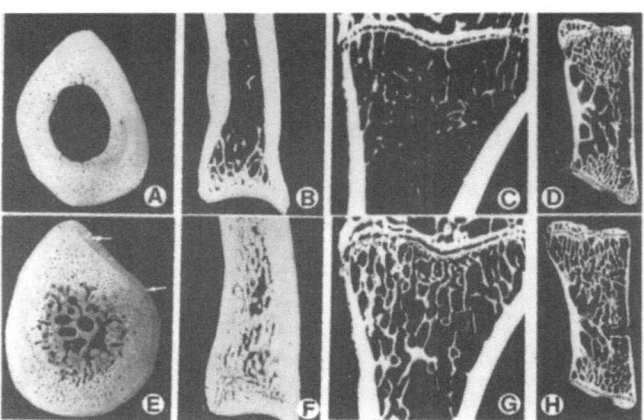

Fig. 2A–H. Extent of PGE$_2$-induced extra bone formation in four select regions of the skeleton after 6 mg PGE$_2$/kg per day for 180 days. PGE$_2$-induced bone formation follows this order: tibial shaft > distal tibia > proximal tibial metaphysis >> lumbar vertebral body. **A-D,** Age-related controls of tibial shaft (**A**), distal tibia (**B**), proximal tibia (**C**), and lumbar vertebral body (**D**). PGE$_2$-treated tibial shaft (**E**), distal tibial metaphysis (**F**), proximal tibial metaphysis (**G**), and lumbar vertebral body (**H**), × 5.5

sociated with a plateau phase of increased bone mass and elevated bone turnover. Indeed its continued administration without adding an antiresorptive agent is currently the only way to preserve the new bone it induces (Jee et al. 1991a; Ke et al. 1992a; Ke and Jee 1992; D–F in Fig. 1c,d).

7.4.2.2 The Consequences of Withdrawal of PGE$_2$ Treatment

Extra bone disappears when treatment is discontinued (Jee et al. 1992a; Ke et al. 1991; Ke and Jee 1992; H and I in Fig. 1c,d)

7.4.2.3 The Distribution of PGE$_2$-Induced Changes in the Rat and the Adult Beagle Dog Skeleton

PGE$_2$ added bone to four sites of the rat skeleton in the following order: TX > DTM > PTM >> LVB (Ito et al. 1993; Fig. 2). The re-

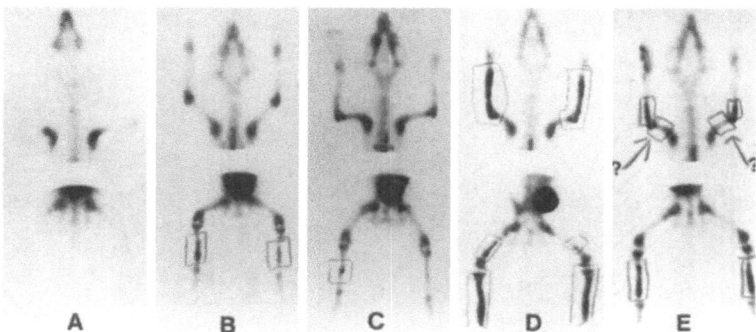

Fig. 3A–E. PGE$_2$-induced high uptake of ^{99m}Tc-bisphosphonate in parts of long bone containing abundant cortical bone in 11-year-old Beagle dogs. **A** Control; **B,C** 0.3 mg/kg per day; **D,E** 1 mg/kg per day. Note the extensive uptake especially in the long bones of the PGE$_2$-treated skeleton (**B–E**) compared to control (**A**)

sponses from these sites of different histomorphometry suggested to us that availability of bone-forming cell progenitors is a key factor that influences the magnitude of the stimulated osteoblastic response. High-turnover bones have fewer progenitor cells available for activation than low-turnover bones. The source of osteogenic cells includes marrow stromal cells, endosteal and intracortical lining cells as well as fat cells (Owen 1980); however, PGE$_2$-induced periosteal woven bone formation in several species indicates adjacent bone marrow is unnecessary. Cortical bone sites with their numerous lining cells in vascular channels are very responsive to osteogenic stimuli (Patt and Maloney 1970). We have frequently been asked how extensive the bone-stimulating effect of PGE$_2$ was on the skeleton as a whole. To answer this we conducted a study in which eleven-year-old Beagle dogs were given 0, 0.3 or 1 mg PGE$_2$/kg per day for 31–59 days and the bone then scanned using ^{99m}Tc-diphosphonate (Fogelman et al. 1978) before sacrifice; high ^{99m}Tc-diphosphonate uptake was seen in long bone cortices (Fig. 3). This further supports previous evidence that the long bone shafts are very responsive to the bone-stimulating effect of PGE$_2$, as seen in humans, dogs, and rats (Lund et al. 1982; Norrdin et al. 1990; Li et al. 1990b; Jee et al. 1985, 1990, 1991a; Tang et al. 1992).

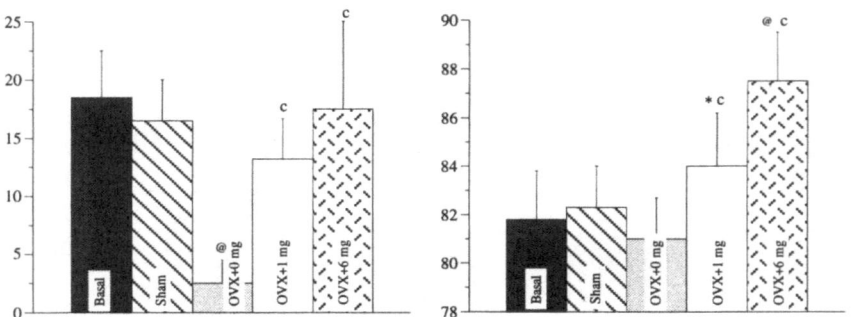

Fig. 4a,b. Prevention of ovariectomy *(OVX)*-induced bone loss by PGE$_2$ (percentages). **a** Proximal tibial metaphysis (PTM) showing 1 and 6 mg PGE$_2$/kg per day for 3 months prevented OVX-induced trabecular bone loss. **b** Tibial shaft showing PGE$_2$ administration added extra cortical bone to OVX rats. OVX induces cancellous bone loss in the PTM **(a)**, but not in the cortical bone of the tibial shaft **(b)**. *$p < 0.05$ and @$p < 0.001$ vs. sham control; $^c p < 0.001$ vs. OVX and 0-mg group

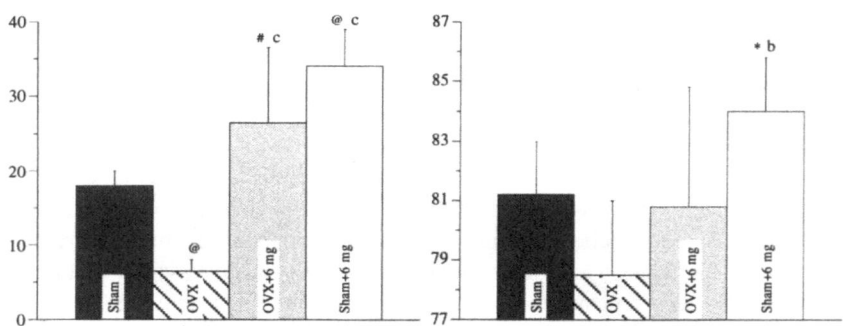

Fig. 5a,b. Restoration and the addition of extra bone by 6 mg PGE$_2$/kg per day for 60 days in the osteopenic proximal tibial metaphysis *(PTM)* and the addition of cortical bone in 4-month post-ovariectomy *(OVX)* rats. **a** PTM showing the percentage of restoration of trabecular bone area above OVX and control levels and the addition of bone above controls in intact female rats. **b** Tibial shaft showing the accumulation of more bone by PGE$_2$ in sham controls than in OVX and aging controls (in percent). *$p < 0.05$, @$p < 0.01$, and #$p < 0.001$ vs. sham controls; $^b p < 0.01$ and $^c p < 0.001$ vs. OVX controls

7.4.2.4 Prevention and Restoration of Osteopenia

PGE$_2$ treatments have prevented ovariectomy- and immobilization-induced bone loss and restored bone to ovariectomy- and immobilization-induced osteopenic bone. Such treatments also added extra bone to increase the outside diameter of long bone cortices and reduced marrow cavity size (Mori et al. 1990, 1992; Jee et al. 1990, 1992a,b; Tang et al. 1992; Akamine et al. 1992; Ke et al. 1992b; Li et al. 1993; Figs. 4,5).

7.4.2.5 The LRM and SM Concepts to Cure Established Osteopenia

Since it is advisable to not use anabolic agents for prolonged periods because of their possible long-term side effects (Lund et al. 1982), cost and the inconvenience of subcutaneous injections, we employed the LRM concept for cancellous bone and the SM concept for cortical bone to reverse existing osteopenia (Tang et al. 1992). The study involved 5.5-month-old female rats ovariectomized for 5 months (lose phase) and then treated with 6 mg PGE$_2$/kg per day for 75 days to restore lost cancellous bone (restore phase). This was followed with withdrawal of PGE$_2$ treatment and treatment with 0,1 or 5 mg/kg twice a week of risedronate, a bisphosphonate, for 60 days (maintain phase). The LRM concept can cure estrogen deficiency-induced cancellous bone osteopenia. When PGE$_2$ treatment was stopped, the PGE$_2$-induced cancellous bone disappeared with no treatment, but 1 or 5 mg risedronate/kg twice a week for 60 days maintained the PGE$_2$-induced increased cancellous bone mass (Fig. 6a,c; Tang et al. 1992) by dramatically reducing bone turnover (Jee et al. 1993). The TX in the LRM experiment lost very little cortical bone so we named the concept for cortical bone supplement and maintain (SM). When PGE$_2$ treatment was stopped and no other treatment given, all the PGE$_2$-induced added bone mass disappeared except the new subperiosteal bone. In contrast, the 1 mg risedronate/kg twice a week maintained some of the subperiosteal, subendocortical, and marrow trabecular bone. With 5 mg risedronate, all the PGE$_2$-induced bone persisted (Fig. 6b,d; Tang et al. 1992) because bone turnover was decreased (Jee et al. 1993).

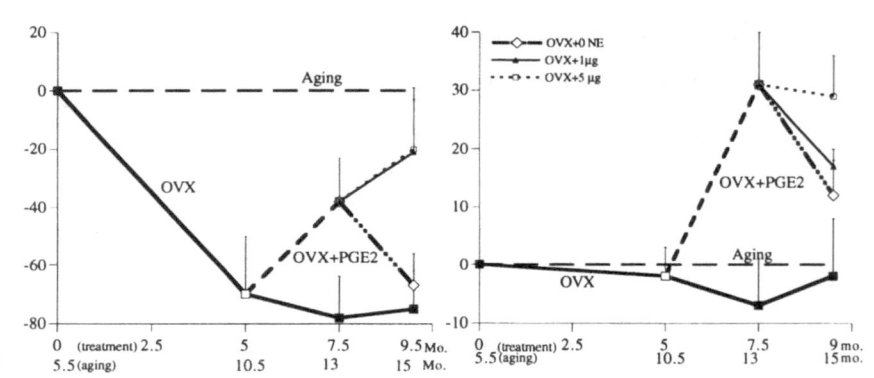

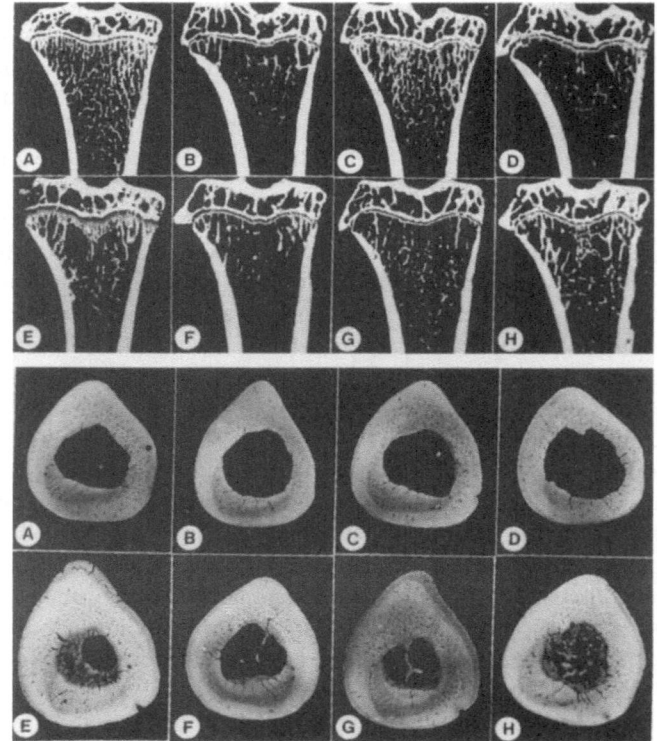

7.4.3 Skeletal Effects of PGE$_2$ on Intact and Castrated Rats

7.4.3.1 Some Select Histomorphometric Features of the Male and Female Skeleton

Much more is known about the bone histomorphometry of the female rat because of the wide use of the ovariectomy model (Kalu 1991; Wronski and Yen 1991). On gross inspection of micoradiographs, we found the that proximal tibia (PT) of male rats is larger with more cortex and has a smaller percentage of metaphyseal trabecular bone than that in females (compare Fig. 7A and Fig. 8A). Histomorphometric analysis showed that the cancellous bone in males has fewer trabeculae with a lower bone formation rate and a higher eroded perimeter and longitudinal growth rate than in females. Examination of the TX confirms that the males possess larger bone (i.e., bone tissue area) with more net cortical bone mass and a larger marrow cavity than females; however, the percentages of cortical and marrow area were the same. In addition, periosteal and endosteal bone formation was elevated. These selected parameters fit more to the paradigm of skeletal adaptation to mechanical usage (SATMU; Frost 1964,1986,1987,1990; Jee et al. 1991b) than to a strict interpretation of the anabolic effects of an androgenic agent. The observation of decreased bone formation rate at the trabecular surface and increased bone formation at the periosteal and endocortical surfaces are in conflict with an interpretation of an overall systemic anabolic response. The male rats being heavier (in terms of body weight) with larger bone with more cortical and less cancellous bone suggests that the cortical bone bears most of the mech-

---

Fig. 6a–d. The lose, restore and maintain (LRM) **(a)** and supplement and maintain (SM) concepts **(b)** and cancellous bone mass (percent trabecular bone area) changes in the proximal tibial metaphysis **(c)** and tibial shaft **(d)** in beginning 5.5-month-old control *(A)*, 5-month post-ovariectomy *(OVX)* (L-phase; *B*), 10.5-month-old control *(C)*, 7.5-month post-OVX *(D)*, 5-month post-OVX plus 2.5 months of PGE$_2$ (R-phase; *E*) and the cessation of PGE$_2$ treatment followed by 0 *(F)*, 1 mg *(G)* or 5 mg *(H)* risedronate twice a week for 2 months (M-phase). The risedronate treatment was able to maintain the cancellous bone mass at the PGE$_2$ level *(G,H)*. The OVX-induced cortical bone loss was small at 5 months post-OVX *(B)*, but the addition of bone is considerable after PGE$_2$ (S-phase, *E*). Unlike in cancellous bone, only 5 mg risedronate was able to maintain the extra bone mass (M-phase, *H*), × 3

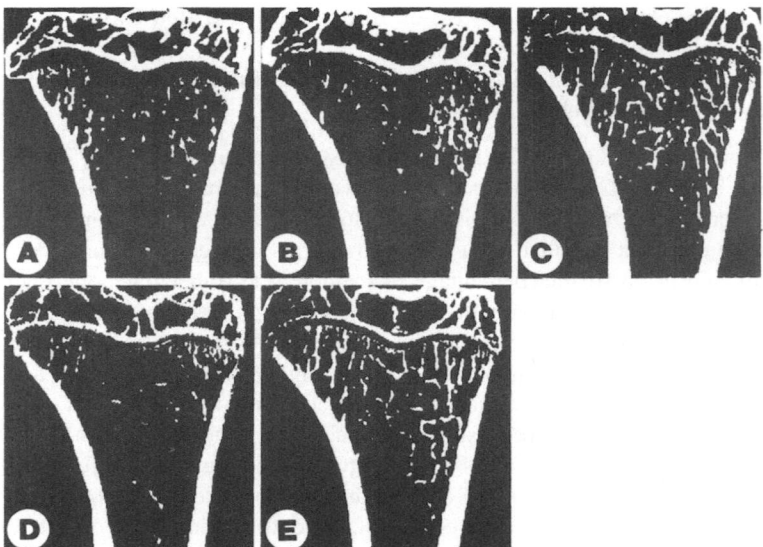

Fig. 7A–E. PGE$_2$ added new bone to the proximal tibial metaphysis (PTM) of intact male and orchiectomized rats. The appearance of the intact male proximal tibia at 3 months beginning control *(A)*, 5 months age-related control *(B)*, 5-month-old treated 2 months previously with PGE$_2$ *(C)*, 5-month-old orchiectomized 2 months previously *(D)*, and 5-month-old orchiectomized and treated with PGE$_2$ for 2 months *(E)*. Note the orchiectomy-induced osteopenia in the PTM *(D)*. PGE$_2$ added new bone to the PTM of both intact *(C)* and orchiectomized rats *(E)*, × 4.5

anical loading, thus freeing the cancellous bone of such duty. If this is true, one can assume the cancellous bone is underloaded and the cortical bone is slightly overloaded compared to the female. The histomorphometric profile of underloading in rats is decreased bone mass with elevated bone resorption (i.e., increased percent eroded perimeter, and decreased trabecular bone formation rate ARF, activation, resorption, formation; Li et al.1990a; Akamine et al. 1992). If the cortex bears most of the skeletal load it is probably slightly overloaded and there would be a higher rate of bone formation at both cortical surfaces (Jee et al. 1991a).

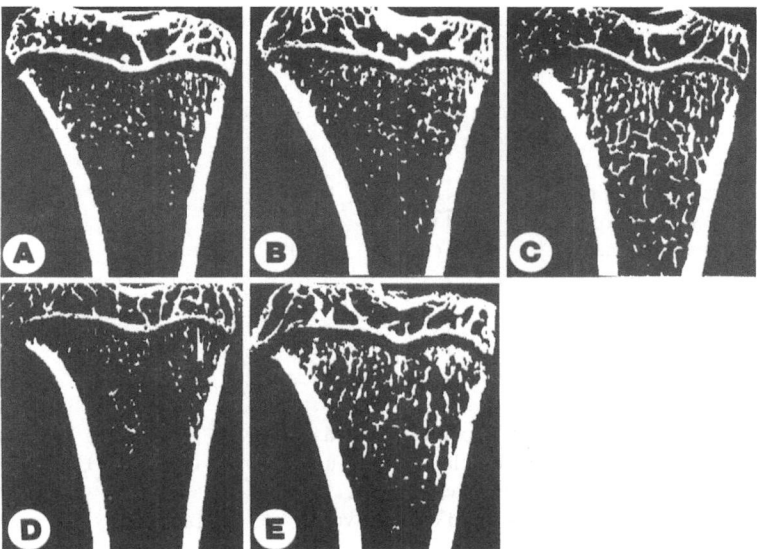

Fig. 8A–E. PGE$_2$ added new bone to the proximal tibial metaphysis (PTM) of intact female and ovariectomized rats. Representative sections of the proximal tibia (PT) from intact female rats of 3-month-old beginning control *(A)*, 5-month-old, age-related control *(B)*, a 5-month-old rat treated with 6 mg PGE$_2$/kg per day for 60 days beginning at age 3 months *(C)*, 5 months of age ovariectomized 2 months previously *(D)*, and a 5-month-old rat ovariectomized and treated with PGE$_2$ at 3 months of age *(E)*. The PGE$_2$ treatment prevented ovariectomy-induced bone loss and added extra bone to the PTM *(E)*. The amount of bone added by PGE$_2$ appears to be about the same amount of cancellous bone regardless whether it involves intact *(C)* or ovariectomized *(E)* rats. In addition, the ovariectomy + PGE$_2$-treated PT *(E)* appears to be larger than all other bones *(A–D)*, × 4.5

7.4.3.2 The Skeletal Responses to Castration

There is general agreement that castration results in cancellous and cortical osteopenia (Saville 1969; Wink and Felts 1980; Schoutens et al. 1984; Verhas et al. 1986; Turner et al. 1989, 1990; Kalu 1991; Wronski and Yen 1991; Wakley and Turner 1991). The bone histomorphometry of the PTM and TX of the intact female and the ovariectomized rats has been well characterized (Mori et al. 1992; Jee et al.

Table 3. Estimate of tissue level changes in orchiectomy and ovariectomized rats

	ORX Young[a]	Older[b]	OVX All ages[c]
Longitudinal bone growth	↓	0	↑
Cancellous bone mass	↓↓	↓↓	↓↓
Trabecular resorption	↑↑↑	↑↑(?)	↑↑↑
Trabecular formation	↑(?)	0(?)	↑
Radial bone growth	↓	0(?)	↑
Cortical bone mass	0	0(?)	↑
Intracortical resorption	↑	↑	↑
Endocortical resorption	↑	↑	↑
Enlarged marrow cavity	↑	↑	↑

ORX, orchiectomized; OVX, ovariectomized

[a]Saville 1969; Wink and Felts 1980; Turner et al. 1989, 1990, 1990d; Schoutens et al. 1984; Wakley et al. 1991.

[b]Verhas et al. 1986; Wink and Felts; University of Utah preliminary observations (our data).

[c]Turner et al. 1987; Wronski et al. 1986; Wronski et al. 1985, 1986; Kalu 1991; Wronski and Yen 1991; Jee et al. 1990, 1991a, 1992a; Ke et al. 1992b; Li et al. 1993; Mori et al. 1990, 1992.

1990; Kalu 1991; Li et al. 1991; Wakley and Turner 1991; Wronski and Yen 1991), but the histomorphometric responses of orchiectomy need to be studied in more depth (Table 3).

The skeletal responses to orchiectomy differ from the ovariectomy-induced responses in that longitudinal and radial bone growth are depressed after orchiectomy while after ovariectomy they are stimulated (Schoutens et al. 1984; Table 3). These functions tend to equalize the size of bones in females and males by depressing bone formation through orchiectomy and stimulating it through ovariectomy. Thus it appeared, in general, sexual dimorphism of the skeleton is abolished by castration (Saville 1969; Wink and Felts 1980; Dahinten and Pucciarelli 1986; Turner et al. 1989, 1990; Wakley and Turner 1991).

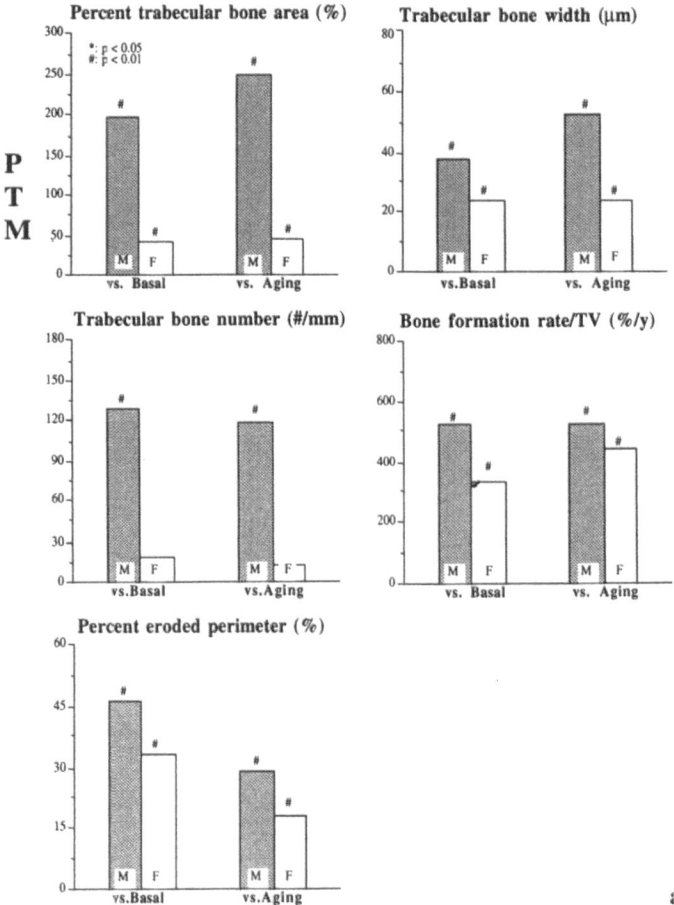

Fig. 9a,b. Effects of PGE$_2$ treatment in the proximal tibial metaphysis (*PTM*, **a**) and tibial shaft (*TX*, **b**) of intact male *(M)* and female *(F)* rats in two different studies: 7-month-old male rats were treated with 6 mg PGE$_2$/kg per day for 60 days and 9-month-old virgin female rats were treated with the same regimen as the male rats. If one assumes that a younger rat should be more responsive to the PGE$_2$, the differences shown above may have been much less evident.

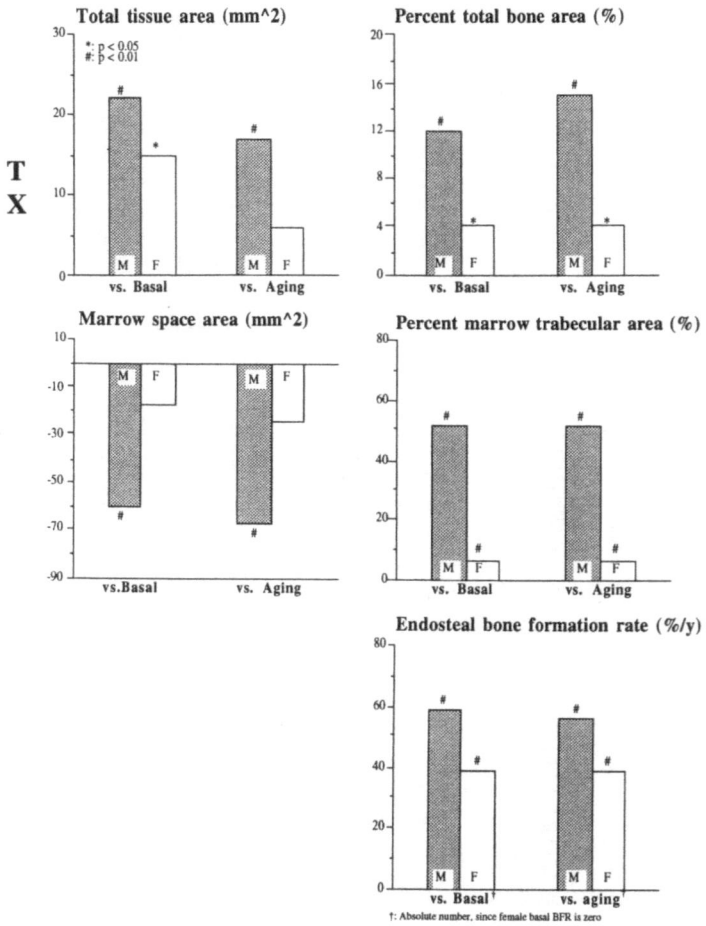

Fig. 9b.

7.4.3.3 PGE$_2$ Adds More Bone in Male than Female Rats

The anabolic effects of PGE$_2$ were markedly greater in the 7-month-old intact male than in the 9-month-old female rats. For example, there was a factor of 4–5 and 3–4 times more new bone added to the PTM (percentage of trabecular bone area; Fig. 9) and TX (percentage total bone area; Fig. 9G) in 7-month-old male and 9-month-old female rats, respectively, treated with PGE$_2$ for 90 days. More impressive is the difference in the stimulation of percentage of marrow trabecular bone in males vs. females, i.e., +51% for males and +6% for females

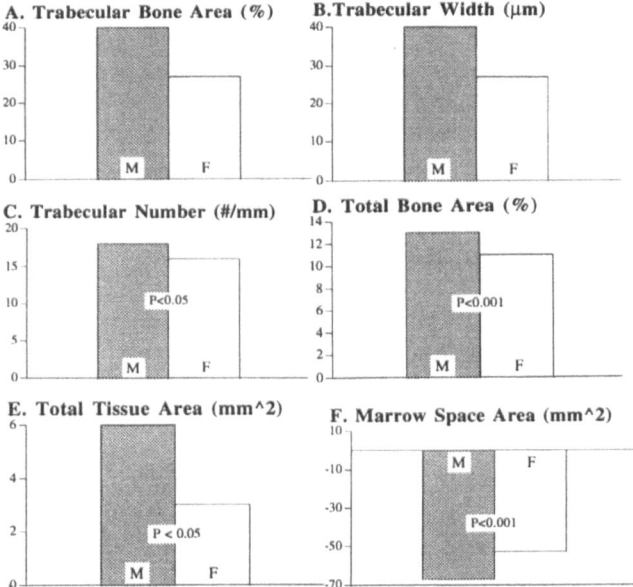

Fig. 10A–F. PGE$_2$ treatment of male and female female rat (PTM) and tibial shaft (TX). Data from intact 3-month-old male and female rats treated with 6 mg PGE$_2$/kg per day for 60 days. Trabecular number (**C**), percent total bone (**D**) and total tissue areas (**E**) were higher, while marrow space areas (**F**) were lower in males than females. The cancellous bone data (**A–C**) are preliminary. The cortical bone data are quite similar to that found in Fig. 9 and led us to a tentative conclusion there is a greater response to PGE$_2$ in male rats

(Fig. 9D). The static morphometric differences between sex in 3-month-old rats was less impressive (Fig. 10). There was no statistically significant difference in most parameters; only total tissue area was an order of magnitude greater in males than females (Fig. 10E). Thus it is possible that there may be no differences in the male or female responses to PGE$_2$ and what was reported in Fig. 9 is simply due to an age effect.

7.4.3.4 PGE$_2$ Stimulates Bone Formation Equally in Castrated Rats

Figure 11 shows that the responses of the orchiectomized and ovariectomized rats to PGE$_2$ were more or less the same in both the cancellous bone of the PTM and the cortical bone of TX. All three (periosteal, endocortical, and trabecular) bone surfaces added extra bone, but the endosteal surface was more reactive than the periosteal surface. The PGE$_2$ added more bone to increase the trabecular width (Fig. 11B) in ovariectomized rats, but fewer numbers of trabeculae (Fig. 11D) than in orchiectomized rates, resulting in about the same amount of cancellous bone (Fig. 11A). The PGE$_2$ added more endocortical bone in the orchiectomized rats to reduce the marrow cavity (Fig. 11F), while the ovariectomized rats added more periosteal bone (Fig. 11E) resulting in approximately the same amount of total cortical bone (Fig. 11A,D).

7.5 Contributions to a Better Understanding of Bone Biology

1. We found that a powerful anabolic agent like PGE$_2$ will stimulate direct bone formation during bone modeling and remodeling.
2. We found that PGE$_2$ shortens the phases of the bone remodeling cycle. It reduces the residence time of the remodeling space and enables the agent to activate new bone turnover sites with a smaller transient bone loss phase that could place an individual at increased fracture risk.
3. We confirm that bone tissue precedes bone marrow in ontogeny.

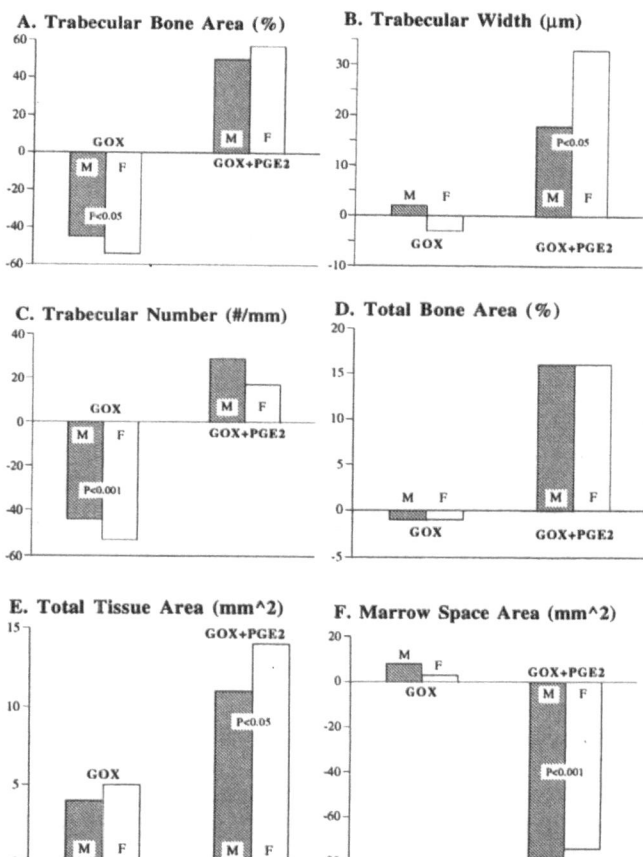

Fig. 11A–F. PGE$_2$ stimulated bone changes in the proximal tibial metaphysis and tibial shaft of castrated *(GOX)* rats. In general, the castration wiped out the sexual dimorphism except for amount of percent trabecular bone area (**A**) and trabecular number (**C**) in castrated male and female rats. PGE$_2$ plus GOX elicited similar PGE$_2$-induces responses, except larger trabecular width (**B**) and total cortical tissue area (**E**) are in favor of the ovariectomized rats. However, the castrated male *(M)* exhibited a greater reduction in marrow space area than the castrated female *(F)*

4. Our studies confirm that cortical bone contains a rich source of osteoblastic progenitors, believed to be the lining cells of the cortical bone vascular channels.
5. We found bone mass increases caused by anabolic agents do not continue indefinitely. One might suggest that they plateau at a new level commensurating with the properties of the agent, dose, and the role of skeletal adaptation to mechanical usage (Frost 1964, 1983, 1986, 1987, 1990; Jee et al. 1991b).
6. Our studies indicate that excess bone induced by an anabolic agent disappears after treatment stops. We postulate that bone adapts to mechanical usage by ridding itself of "extra" bone that keeps strain "too low," or adding "extra" bone when strain is "too high," to maintain constant strain within the physiologic window (Frost 1964, 1983, 1986, 1987; 1990; Turner 1992).
7. In our studies we discovered that we could outsmart the skeleton by using the LRM concept to restore and maintain PGE$_2$ added cancellous bone and the SM concept to build up bone.
8. We found PGE$_2$ will stimulate periosteal bone formation and consequently increase the diameter of the cortex and its strength. The engineers tell us that the strength of a given mass can be increased by increasing the diameter of the shaft while making its wall proportionally thinner (Ruff and Hayes 1983; Burr and Martin 1983; Martin and Burr 1989; Turner 1991).
9. We found that PGE$_2$ depresses longitudinal bone growth in male but stimulates it in female ovariectomized and immobilized rats. This is an area of research that warrants more study.
10. We confirm that the DTM closes at 3 months and a few months later contains a metaphysis with fewer but thicker trabeculae and lower turnover. Studies of the adult metaphysis could improve our understanding of the responses of this low turnover site to bone-seeking agents.

7.6 Concluding Remarks

We do not advocate employing PGE$_2$ itself as a therapeutic agent for the stimulation of bone formation due to its side effects, but we do believe our studies have improved our understanding of the mode of anabolic agents in bone. These studies should also encourage the use of such agents in preclinical studies and their developmentOur results of the effects of PGE$_2$ on cancellous and cortical bone by histomorphometry are invaluable in that they bridge the gap between cellular and molecular biology and clinical investigations. Furthermore, the preclinical studies employing the various rat osteopenic models can be so comprehensive that there is no need to do large animal efficacy studies (except for toxicity and dose studies) before proceeding to the clinic. We further emphasize that it is important to do long-term studies to establish the status of the new steady state as well as to determine whether the agent is truly a bone-forming agent.

Currently, many investigators are suggesting that a given agent is anabolic on the basis of transient data. We have also studied the cortical bone responses along with the more popular cancellous bone response, which most laboratories have avoided, because an increasing number of us have become aware that the cortical bone plays a larger role in the prevention of fractures than cancellous bone (Mazess 1990). There is increasing evidence that therapies for excessive skeletal fragility should be directed toward increasing cortical bone mass, with emphasis on increasing subperiosteal bone. Along this line, more biomechanical testing should be performed. I have always maintained that there is a need to test cancellous bone biomechanically, but now I advocate more extensive testing of cortical bone sites.

There is a paucity of data on the male rat skeleton and the orchiectomy-induced osteopenia rat. We plan to do our part and hope to soon complete our histomorphometric study of the growing and aged male rat skeleton. This information should allow for better planning of male skeletal studies.

Lastly, there are many problems with anabolic agents needing answers that can be studied in the rat. The most obvious is the role of skeletal adaptation to mechanical usage in regulating bone mass. Does it limit the amount of bone formed by a bone-forming agent at a given site? How much mechanical loading is needed to maintain bone? Are there agents that will maintain the "extra" bone without reducing activation frequency that places osteoporotics at more risk because of the possible lack of repair of microdamage.

Acknowledgments. The senior author wishes to thank the following postdoctoral fellows for their pioneering and continued efforts to better understanding of the skeletal biology and the action of PGE$_2$ in the rat. They include J. Inoue, K. Ueno, T. Haba, Y. Furata, T. Tanisawa, S. Mori, T. Akamine, S. Chan, L.Y. Tang, M.M. Chen, H. Ito, Q.Q. Zeng and H. Ito.

References

Akamine T, Jee WSS, Ke HZ, Li XJ, Lin BY (1992) PGE$_2$ prevents disuse-induced cancellous bone loss and adds extra bone to immobilized bones. Bone 13:11–12

Burr DB, Martin RB (1983) The effects of composition, structure and age on the torsional properties of the human radius. J Biomechanics 16:603–608

Compston JE, Mellish RWE, Garrahan NJ (1987) Age-related changes in iliac crest trabecular microanatomic bone structure in man. Bone 8:289–292

Dahinten SL, Pucciarelli HM (1986) Variations in sexual dimorphism in the skulls of rats subjected to malnutrition, castration and treatment with gonadal hormones. Am J Physical Anthropol 71:63–67

Dawson AB (1925) The age order of epiphyseal union in long bones of albino rats. Anat Rec 31:1–17

deWinter FR, Steendijk R (1975) The effect of a low-calcium diet in lactating rats; observations on the rapid development and repair of osteoporosis. Calcif Tissue Res 17:303–316

Fogelman I, Bessent RG, Turner JG (1978) The use of whole body retention of Tc-99m diphosphonate in the diagnosis of metabolic bone disease. J Nucl Med 19:270–274

Foresta C, Zannatta GP, Busnardo B, Scanelli G, Scandellari C (1985) Testosterone and calcitonin plasma levels in hypogonadal osteoporotic young men. J Endocrinol Invest 8:377–379

Frost HM (1964) The laws of bone structure. Thomas, Springfield

Frost HM (1977) A method of analysis of trabecular bone dynamics. In: Meunier PJ (ed) Bone histomorphometry. Armour-Montagu, Paris, pp 445–476

Frost HM (1983) The minimum effective strain: a determinant of bone architecture. Clin Orthop 175:286–292

Frost HM (1986) Intermediary organization of the skeleton, vols I and II. CRC, Boca Raton

Frost HM (1987) The mechanostat: a proposed pathogenic mechanism of osteoporoses and the bone mass effects of mechanical and nonmechanical agents. Bone Miner 2:73–85

Frost HM (1990) Skeletal structural adaptations to mechanical usage (SATMU): 1. Redefining Wolff's law: the bone modeling problem. 2. The remodeling problem. Anat Rec 226:403–422

Frost HM, Jee WSS (1992) On the rat model of human osteopenias and osteoporoses. Bone Miner 18:227–236

Ito H, Ke HZ, Jee WSS, Sakou T (1993) Anabolic responses of an adult cancellous bone site to prostaglandin E_2 in the rat. Bone Miner 21:219–236

Jackson JA, Kleerekoper M, Parfitt AM, Rao DS, Villanueva AR, Frame B. (1987) Bone histomorphometry in hypogonadal and eugonadal men with spinal osteoporosis. 65:53–58

Jee WSS (1991) The aged rat model for bone biology studies. Cells Materials [Suppl 1]:1–192

Jee WSS, Inoue J, Jee KW, Haba T (1983) Histomorphometric assay of the growing long bone. In: Takahashi H (ed) Handbook of bone morphology. Nishimura, Niigata City, Japan, pp 101–122

Jee WSS, Ueno K, Deng YP, Woodbury DM (1985) The effects of PGE_2 in growing rats: increased metaphyseal hard tissue and cortico-endosteal bone formation. Calcif Tissue Int 37:148–157

Jee WSS, Ueno K, Kimmel DB, Woodbury DM, Price P, Woodbury LA (1987) The role of bone cells in increasing metaphyseal hard tissue in rapidly growing rats treated with PGE_2. Bone 8:171–178

Jee WSS, Mori S, Li XJ, Chan S (1990) PGE_2 enhances cortical bone mass and activates intracortical bone remodeling in intact and ovariectomized female rats. Bone 11:253–266

Jee WSS, Ke HZ, Li XJ (1991a) Long term anabolic effects of PGE_2 on tibial diaphyseal bone in male rats. Bone Miner 15:33–55

Jee WSS, Li XJ, Ke HZ (1991b) The skeletal adaptation to mechanical usage in the rat. Cells Materials [Suppl 1]:131–142

Jee WSS, Ke HZ, Li XJ (1992a) Loss of PGE_2-induced cortical bone after its withdrawal in rats. Bone Miner 17:31–47

Jee WSS, Akamine T, Ke HZ, Li XJ, Tang LY, Qeng QQ (1992b) PGE_2 prevents disuse-induced cortical bone loss. Bone 13:153–159

Jee WSS, Tang L, Ke HZ, Setterberg RB, Kimmel DB (1993) Maintaining restored bone with bisphosphonate in the ovariectomized rat skeleton: dynamic histomorphometry of changes in bone mass. Bone 14:481–485

Johnston CC (1985) Studies on prevention of age-related bone loss. In: Peck WA (ed) Bone and mineral research/3. Elsevier, Amsterdam, pp 233–257

Kalu DN (1991) The ovariectomized rat model of postmenopausal bone loss. Bone Miner 15:175–192

Ke HZ, Jee WSS (1992) Effects of daily administration of PGE_2 and its withdrawal on the lumbar vertebral bodies in male rats. Anat Rec 234:172–182

Ke HZ, Jee WSS, Li XJ (1991) Partial loss of anabolic effect of prostaglandin E_2 on bone after its withdrawal in rats. Bone 12:173–183

Ke HZ, Jee WSS, Mori S, Li XJ, Kimmel DB (1992a) Effects of long-term daily administration of prostaglandin E_2 on maintaining elevated proximal tibial metaphyseal cancellous bone mass in male rats. Calcif Tissue Int 50:245–252

Ke HZ, Li M, Jee WSS (1992b) Prostaglandin$_2$ prevents ovariectomy-induced cancellous bone loss in rats. Bone Miner 9:45–62

Kimmel DB (1991) The oophorectomized beagle as an experimental model for estrogen-depletion bone loss in adult human. Cells Materials [Suppl 1]:75–84

Li XJ, Jee WSS, Chow SY, Woodbury DM (1990a) Adaptation of cancellous bone to aging and immobilization in the rat. Anat Rec 227:12–24

Li XJ, Jee WSS, Li YL, Patterson-Buckendahl P (1990b) Transient effects of subcutaneously administered PGE_2 on cancellous and cortical bone in young adult dogs. Bone 11:353–364

Li XJ, Jee WSS, Ke HZ, Mori S, Akamine T (1991) Age related changes of cancellous and cortical bone histomorphometry in female Sprague-Dawley rats. Cells Materials [Suppl 1]:25–35

Li M, Jee WSS, Ke HZ, Liang XG, Lin BY, Ma YF, Setterberg RB (1993) Prostaglandin E_2 restores cancellous bone to immobilized limb and adds bone to overloaded limb in right hindlimb immobilization rats. Bone 14:283–288

Lindgren JU (1976) The effect of thyroparathyroidectomy on development of disuse osteoporosis in adult rats. Clin Orthop 118:251–255

Lund JE, Brown WP, Tregerman L (1982) The toxicology of PGE_2. In: Wu KK, Rossi EC (eds) Prostaglandin in clinical medicine: cardiovascular and thrombotic disorders. Yearbook Medical, New York, pp 93–109

Martin RB, Burr DB (1989) Structure, function and adaptation of compact bone. Raven, New York, pp 230–233

Mazess RB (1990) Fracture risk: a role for compact bone. Calcif Tissue Int 47:191–193

Mori S, Jee WSS, Li XJ, Chan S, Kimmel DB (1990) Effects of PGE_2 on production of new cancellous bone in the axial skeleton of ovariectomized rats. Bone 11:103–113

Mori, S, Jee WSS, Li XJ (1992) Production of new trabecular bone in osteopenic ovariectomized rats. Calcif Tissue Int 50:80–87

Norrdin RW, Jee WSS, High WB (1990) The role of prostaglandins in bone in vivo. Prostaglandins Leukot Essent Fatty Acids 41:139–149

Owen M (1980) The origin of bone cells in the postnatal organism. Arthritis Rheum 23:1073–1079

Parfitt AM (1990) Bone-forming cells in clinical condition. In: Hall BK (ed) The osteoblast and osteocyte. Telford, Caldwell, NJ, pp 351–429

Parfitt AM, Drezner MK, Glorieux FH, Kanis JA, Malluche H, Meunier PJ, Ott SM, Recker RR (1987) Bone histomorphometry: standardization of nomenclature, symbols and units. J Bone Miner Res 2:595–610

Patt HM, Maloney MA (1970) Reconstitution of bone marrow in a depleted medullary cavity. In: Stohlman F (ed) Hemopoietic cellular proliferation. Grune and Stratton, New York, pp 56–66

Ruff CB, Hayes WC (1983) Cross-sectional geometry of Pecos Pueblo femora and tibiae: a biomechanical investigation. I. Method and general patterns of variation. Am J Phys Anthrop 60:359–381

Ruth EB (1953) An experimental study of the haversian-type vascular channels. Anat Rec 112:429–455

Saville P (1969) Changes in skeletal mass and fragility with castration in the rat: a model of osteoporosis. Am Geriatric Soc 17:155–166

Schoutens A, Verhas M, L'Hermitie-Baleriaux M, L'Hermitie M, Verschaeren A, Dourov N, Mone M, Heilporn A, Tricot A (1984) Growth and bone haemodynamic responses to castration in male rats. Reversibility by testosterone. Acta Endocrinol 107:428–432

Tang LY, Jee WSS, Ke HZ, Kimmel DB (1992) Restoring and maintaining bone in osteopenic female rat skeleton: I. Changes in bone mass and structure. J Bone Miner Res 7:1093–1104

Thorngren KG, Hansson LI (1973) Cell kinetics and morphology of the growth plate in the normal and hypophysectomized rat. Calcif Tissue Res 13:113–129

Turner CH (1991) Toward a cure for osteoporosis: reversal of excessive bone fragility. Osteoporosis Int 2:12–19

Turner CH (1992) Editorial: functional determinants of bone structure: beyond Wolff's law of bone transformation. Bone 13:403–409

Turner RT, Vandersteenhoven JJ, Bell NH (1987) The effects of ovariectomy and 17β-estradiol on bone histomorphometry in growing rats. J Bone Miner Res 2:115-122

Turner RT, Hannon KS, DeMers LM, Bell NH (1989) Differential effects of gonadal function on bone histomorphometry in male and female rats. J Bone Miner Res 4:557–563

Turner RT, Wakley GK, Hannon KS (1990) Differential effects of androgens on cortical bone histomorphometry in gonadectomized male and female rats. J Orthop Res 8:612–617

Ueono K, Haba T, Woodbury D, Price P, Anderson R, Jee WSS (1985) The effects of prostaglandin E_2 in rapidly growing rats depressed longitudinal and radial growth and increased metaphyseal hard tissue mass. Bone 6:79–86

Verhas M, Schoutens A, L'Hermitie-Baleriaux M, Dourov N, Verschaeren A, Mone M, Heilporn A (1986) The effect of orchiectomy on bone metabolism in aging rats. Calcif Tissue Int 39:74–77

Wakley GK, Turner RT (1991) Sex steroids and the regulation of bone volume in the rat. Cells Materials [Suppl] 1:85–91

Wakley GK, Schutte HD Jr, Hannon KS, Turner RT (1991) Androgen treatment prevents osteopenia of cancellous bone in the orchiectomized rat. J Bone Min Res 6:325–330

Waynford HB (1980) Experimental and surgical technique in the rat. Academic Press, New York

Wink G, Felts W (1980) Effects of castration on the bone structure of male rats: a model of osteoporosis. Calcif Tissue Int 32:77–82

Wronski TJ, Lowry PL, Walsh C, Ignaszewski LA (1985) Skeletal alterations in ovariectomized rats. Calcif Tissue Int. 37:324-328

Wronski TJ, Walsh C, Ignaszewski LA (1986) Histologic evidence for oseopenia and increased bone turnover in ovarietomized rats. Bone 7:119-123

Wronski TJ, Yen CF (1991) The ovariectomized rat as an animal model for postmenopausal bone loss. Cells Materials [Suppl 1]:69–74

8 Animal Models for Osteoporosis

Hartmut H. Malluche and Janet B. Rodgers

8.1 Introduction

Osteoporosis, its clinical sequela manifested by fractures, has become a worldwide health and socioeconomic problem of major proportions. Presently an estimated 75 million people in the United States, Europe, and Japan are afflicted with this disease. The prevalence of osteoporosis has increased in western countries with a growing incidence of vertebral and hip fractures together with a decrease in the age at which they occur. Direct and indirect costs are calculated to be approximately $10 billion per year. Annually, osteoporosis is responsible for at least 1.3 million fractures, 30 000 deaths and long-term domiciliary care in 50% of the survivors in the United States alone.

Factors that play a role include aging and estrogen deficiency. Patient studies are limited by variables such as peak bone mass at skeletal maturity and genetic factors that affect bone remodeling and balance.

Studies are further confounded by additional risk factors such as inadequate calcium nutrition, cigarette smoking, alcohol abuse, sedentary lifestyles, and medication. Because of these numerous variables, animal models are an essential to general research efforts. These models have also proven to be critical to the adequate evaluation of drugs. However, all animal models have limitations and therefore many factors have to be considered in their selection. These factors are discussed in this paper.

Perhaps most obviously, it needs to be understood that osteoporosis is not a homogeneous disorder. While osteoporosis occurring right after menopause (type 1) is easier to study with surgical removal of ovaries or endocrine manipulation of the ovarian function through administration of gonadotropin-releasing hormone (GNRH), an estrogen-releasing hormone, osteoporosis seen at old age (type 2) is both costly and difficult to reproduce in animals. Still, despite constraints, advances in research will most likely continue to depend on the use of animal models. This leaves us with the responsibility of weighing the costs and benefits when considering the effectiveness of various animal models.

Kalu expanded on a definition of an animal model (Wessler 1976) as a living animal in which spontaneous or induced bone loss due to ovarian hormone deficiency can be studied and in which the characteristics of the bone loss and its sequelae resemble those found in postmenopausal women in one or more respects (Kalu 1991). There are at least three characteristics of an animal model: (1) convenience, (2) relevance (comparability to the human condition), and (3) appropriateness (a complex of other factors that make a given species the best for studying a particular phenomenon). While each animal model presented for discussion has certain drawbacks, the use of these models offers a significant contribution to a better understanding of the effects of estrogen-induced bone loss on the human skeleton.

8.2 Primates Versus Nonprimates

While rodents, dogs, sheep, and swine are among the most commonly used nonprimate animal models, in terms of relevance, i.e., comparability to the human condition, nonhuman primates are the most ideal

choice of an animal model. They have a similar physiological menopause to humans, unlike nonprimates who do not undergo menopause.
However, nonhuman primates are prohibitive in price. Their psychological disposition alone makes handling and maintenance costly.
Also, housing requirements in the U.S. are strict. New regulations
mandate environmental enrichment and more space. Additionally,
handling of primates can be dangerous due to the zoonotic diseases
they may carry. Because of this they are usually anesthetized before
even minor blood sampling or routine health testing is performed.
These economic factors have severely limited the use of the nonhuman
primate model.

Despite the high costs associated with the use of nonhuman primates, the nonhuman primate model has certain indisputable advantages in comparing animal subjects to human bone. Many primates
maintain an upright body posture, therefore their bone biomechanics
are more similar to that of human. Even more importantly, the estrous
cycles and menopause in some primates is closest to that of human females, with similar longevity. Finally, their dietary requirements are
more like humans than any other domesticated species. While a few investigators have examined nonhuman primates as models, restrictions
to their use will continue to render them an impractical alternative.
This leaves nonprimate animal models as the most feasible alternative
for the majority of studies.

Compared to primates, nonprimates appear less than ideal. As
quadrupeds, nonprimates may have different bone biomechanical
characteristics than bipeds (Turner 1991). Mechanical forces that act
upon the skeleton govern regulation of bone mass and three-dimen-
sional structure. Loss of estrogen may produce lasting effects upon
bone structure by modulating the "set point" for bone adaptation.
However, while there is no domestic animal model for bone loss after
cessation of ovarian function which perfectly replicates the human
condition, experimental models have yielded and continue to provide
useful information.

8.3 Rodents

Laboratory rats have served diverse research needs more than any other animal species. Rats have certain practical advantages in bone experiments. First of all, they are less costly than larger laboratory animals. They are easy to care for and house. Since their life span averages 3–4 years it is easier to study aging changes in rats. They are also genetically well defined. Mutants exhibiting specific disease processes have been well described in the literature (Kohn et al. 1984). Transgenic mice have been used more frequently in laboratory experiments involving genetic manipulation.

Despite these apparent advantages, rats may represent a false economy, particularly when research requires large numbers of experimental groups. Longitudinal studies cannot be performed if repetitive bone sampling and large blood samples prove necessary to meet research needs. Pharmacokinetics often differ from humans. This means that pharmacologic agents and experimental compounds must be carefully administered with these differences in mind. This may require separate control and experimental groups. This not only adds higher costs but it confounds research efforts with the increased risk found in adding on another uncontrolled variable.

Generally, drawbacks to experimental use of rats include their small body size, short life span, low blood volume, and high basal metabolic rate. Their estrous cycle is short (12 h of estrous recurring every 4–5 days throughout the year). Additionally the rat is a nocturnal animal. Therefore its diurnal rhythms must be taken into account in experiments.

However, rats are a familiar subject and the use of laboratory rats has proven to be a valuable aid to research. Much is known about mineral metabolism in rats as a result of extensive research in several scientific disciplines. Metabolically, serum levels of calcium and phosphorus have been measured in many individual strains and summarized according to strain, gender, age, and other factors. This information can be provided by the vendor. Serum phosphorus levels vary widely depending on the method of analysis as well as the above-mentioned factors; for female rats pooled over age and strain, a mean phosphorus level has been reported to be 5.75 ± 2.58 mg% (Loeb and Quimby 1989). Higher normative levels have also been reported (i.e.,

7.08 ± 1.19 mg/dl; Capen and Rosol 1989; Loeb and Quimby 1989). For calcium, a range of 10–13 mg/dl (mean 11.75 ± 089) might be used. (Loeb and Quimby 1989).

Two general models of ovariectomy-induced bone loss have been published in rats: those using mature animals and those using aged animals. A review of the effects of aging upon the rat model was published by Kalu (Kalu 1991). Erben et al. ovariectomized Fischer 344 rats at 10 weeks of age and followed them for 16 weeks postoperatively (Erben et al. 1992). Other experimenters have used a model with Sprague-Dawley rats which were ovariectomized at 90 days of age and followed for variable periods up to 540 days (Kimmel and Wronski 1990; Wronski et al. 1987).

In some experiments results in obese rats of the same strain were obtained. In contrast to these mature animals, others have studied aged rats (Faugere et al. 1982). In early studies, Holtzman rats were ovariectomized at 50 weeks of age. Starting 38 weeks later, treatments with calcitriol were instituted. At 102 weeks of age, histomorphometric study showed decreased cancellous and cortical bone mass and a decreased ratio between mineralized and nonmineralized bone in untreated ovariectomized rats. In calcitriol-treated, ovariectomized rats the bone loss was partially corrected, and the mineralized to nonmineralized bone ratio was normal, as was the restored growth plate width (Faugere et al. 1986). Ibbotson et al. (1992) studied rats ovariectomized at 1 year of age beginning their studies of parathyroid hormone (PTH) and insulin-like growth factor I (IGF-I) at 2 years of age. A recent conference concluded that age is a major factor in the outcome of bone loss experiments and that the growth curve of female rats slows dramatically after 1 year of age.

As long as the rat's limitations are taken into account, its contribution to bone research is of value. For instance, the rat model most commonly employed in bone experiments lack Haversian systems like those of larger mammals, and systemic internal remodeling of cortical and cancellous is dissimilar. In our experience, individual variation in bone parameter such as cancellous bone volume can be problematic. We have also observed that vitamin D-deficient rats exhibit increased serum phosphorus levels, whereas dogs and humans do not.

If skeletal maturity is a requirement of the study, aged rats must be used. Kalu measured femoral length, weight, density, and calcium levels in female Wistar rats and determined that by 12 months of age the skeleton has stabilized at a level close to that of senescent (> 2 years) rats (Kalu et al. 1989). Chow et al. (1993) claim in this issue that epiphyseal closure of the tibia occurs at around 2 years of age.

Body weight cannot be cited as an indicator of age, since different strains have different growth curves. For example, a 250g female Wisconsin rat (Crl: (WI)BR) may be 53–62 days of age, while a Long-Evans rat of the same weight and gender (Crl: (LE)BR) may be 91–107 days old.

Anatomically, the rat has a single pair of parathyroid glands located close to the thyroid (Capen and Rosol 1989).

The National Research Council (NRC) has determined minimum nutritional requirements for many species, including the rat. Although standardized rations are quite adequate for many experiments, some contamination of feed components with synthetic estrogens, mycotoxins, heavy metals, and insecticides may occur and have undesirable effects on experimental results. For this reason the experimenter should question the animal care facility about the standard diet and possibly request that the rats be fed a chemically defined or purified diet.

8.4 Dogs

Next to rodents, dogs are the most frequently used animal model . Unlike rats, they possess Haversian systems and they have a high level of internal remodeling activity similar to humans. Furthermore, their calcium and phosphorus metabolism has been well documented. However, dogs are seasonally monoestrous, meaning that they exhibit only a single estrous period on a seasonal basis (usually once per year). This obvious departure from the human reproductive physiology needs to be considered by the experimenter. If deemed necessary, the dogs' estrous cycles may be synchronized during a long-term study to ascertain comparability of baseline values.

It should be noted that although random-bred dogs are available from a variety of sources, they come from unknown backgrounds and they should not be used in bone metabolism research. Purpose-bred dogs have a consistent background of proper husbandry and veterinary care. They have had routine health checkups, appropriate vaccinations, and parasite treatment schedules. Additionally, they have had a proper diet and sufficient care. Any undesirable or defective animals are culled from the colony.

Beagles are the most commonly used purpose-bred dogs for research for several reasons. The breed is of moderate body size. They adapt well to confinement and they possess a gentle easygoing nature.

Our research group established the beagle model for estrogen-deficiency induced bone loss (Malluche et al. 1986, 1988; Faugere et al. 1990). The dogs have closed epiphyses (at least 4–6 years old) when the first iliac crest bone biopsy is taken, followed by ovariohysterectomy. An initiation phase of bone loss occurs within 1 month after surgery, consisting of an increase in trabecular separation accompanied by a dramatic decrease in cancellous bone volume. These changes can be prevented by administration of a bisphosphonate starting at the time of ovariohysterectomy (Monier-Faugere et al. 1992). The second, or maintenance, phase is related to osteoblastic insufficiency, which is partially correctable by administration of $1,25(OH)_2D_3$ (Malluche et al. 1990). Nakamura et al. (1992) ovariectomized 2- and 3-year-old beagles and monitored serum chemistries, bone gla-protein, urinary hydroxyproline, and vitamin D metabolites for 31 months postoperatively. Single-photon absorptiometry was performed on L6-7 and one femur after the animals were killed and an 8-mm trephoned core of ilium was subjected to histomorphometry. Their results indicated that administration of $24R,25(OH)_2D_3$ inhibited the increase in bone turnover and prevented reduction of cancellous bone mass. Shen et al. (1992) were unsuccessful in producing significant measurable changes in histomorphometric parameters in a similar model. This lack of success may possibly have been due to limitations in the ages used, number of dogs per group, and differences in how bone biopsies were obtained.

It is critically important to standardize methods for obtaining and evaluating bone biopsies, both within and between research institutions. Following ovariohysterectomy in the dog, the cortices of the ilium may move together soon after the surgery, as deep cancellous bone is lost. The thick, subcortical cancellous bone trabeculae are typically thicker, a feature which must be taken into account when making histomorphometric measurements. Deep cancellous, subcortical cancellous, and cortical bone should be evaluated separately. In order to include all three compartments and to have sufficient bone material for analysis, biopsies may not be taken using a trocar designed for transversal sampling. We use a Stryker bone saw and obtain a large wedge of bone from a standard site, employing aseptic surgical technique. Repeat biopsies may be taken from the contralateral ilium.

The dog has two pairs of parathyroids (an external and internal pair) with separate blood supplies (Capen and Rosol 1989). Endocrine abnormalities are observed in some older animals, particularly of the thyroid and parathyroid glands. Therefore thyroid screening should be performed routinely upon all dogs in bone metabolism studies to rule out this variable.

Diet is a particular maintenance concern. Dogs are omnivorous animals. They possess a gastrointestinal tract which is more like the human's than any rodent species. Confusion over what constitutes a proper diet in the past has led to the feeding of all-meat diets. This has caused nutritional secondary hyperparathyroidism due to the low calcium content and unfavorable Ca:P ratio for such diets (Fraser 1986). Some commercial pet diets may still contain more calcium and phosphorus than are optimum for health. As with rodents, the investigator must be aware of canine dietary requirements established by the NRC and discuss any potential conflicts in bone metabolism experiments with animal care staff prior to beginning any studies.

Dietary calcium requirements are higher in dogs than humans, although they are lower than in most other domestic animal species. NRC requirements state that for calcium 120 mg/kg body wt per day are best, for phosphorus 90 mg/kg per day, and that the dietary calcium ratio should be between 1.2–1.4:1 (Capen and Rosol 1989). Furthermore, the dog has a lower intestinal fractional calcium absorption than humans (Lopez-Hilker et al. 1986).

8.5 Sheep

With the help of experienced technical personnel, sheep are easy to handle. Additionally, large blood samples are easy to obtain and repetitive bone biopsies are not difficult if the appropriate surgical facilities are available. The sheep is a seasonally polyestrous animal. and it experiences several estrous cycles during its breeding season. This season can vary depending on the breed of sheep, the day length, or indoor lighting conditions. The estrous cycle is 14–21 days in length, although this varies with the breed.

However, the availability of sheep depends on where the research institution is located (Pastoureau et al. 1989). Due to their large body size (approximately 60–100 kg) sheep require more space to house and specialized handling or restraint equipment may be necessary.

Park, Turner, and coworkers have begun to work with ovariectomized sheep as a model for bone loss (Park et al. 1992). They reported that after 6 months ovariectomized sheep demonstrated selective trabecular loss when compared with either young (3–4 years) or old (8–9 years) ewes. The same group reported that aged sheep display bone loss at a caclulated rate of 6% per decade, when compared with younger animals.

The sheep is a herbivorous ruminant animal, which engenders potential differences in metabolism of vitamins and dietary minerals. Consultation with veterinarians should be sought if sheep are to be compared to humans in bone experiments.

Mature sheep require 60 mg/day calcium and 55 mg/day phosphorus. As in other species the calcium to phosphous ratio should not be less than 1–1.5:1 (Capen and Rosol 1989). The sheep, like the dog, has two pairs of parathyroids. Sheep calcitonin has been sequenced, with 13 of its 32 amino acids homologous to the human type.

8.6 Swine

Like humans the pig exhibits continuous estrous cycles and the length of each cycle (18–20 days) is also similar. It has been reported that sows housed without the presence of boars exhibit a later onset of puberty (Mavrogenis et al. 1976; Pond and Haupt 1978). We have demonstrated that Yucatan minipigs kept in conventional biomedical facilities have regular estrous cycles and reach puberty at approximately 5–6 months of age, in the absence of males (Rodgers et al. 1993). Ovariohysterectomy is more difficult to perform than in the dog, because the blood supply to the uterus is more friable.

The pig has lamellar bone and trabecular and cortical remodeling which are similar to that of humans. Induction of fluorosis has effects that can be compared closely to humans, although there is a discrepancy between bone mass and strength changes (Mosekilde et al. 1987). Mosekilde et al. also reports that pigs have a higher bone mass and denser trabecular network than humans. The reasons for this are unknown (Mosekilde et al. 1987).

The commercial farm pig and the miniature pig are two distinct types of pig used in biomedical research. The obvious disadvantage of farm pigs is that they grow to an adult body weight of approximately 150 kg while miniature swine rarely exceed 60 kg at maturity. Therefore a 20-kg farm pig will be much younger in age than a minipig of the same weight. While miniature swine are more expensive to purchase and may be unavailable in some areas, they require less housing space than farm swine and no special handling equipment is needed.

There are several breeds of miniature swine available, originating from different stocks. The Sinclair, Hanford, and Hormel minipigs originated from wild pigs that were crossbred with farm pigs. They were developed in the 1950s in the U.S., partly because their skin made them valuable models for studies of radiation burns. The Yucatan minipig, originating from Mexico, has a gentle disposition. When handled by competent technicians they can be trained to accept tasks such as using a treadmill. In recent years the Vietnamese Pot Bellied pig, originating from India and Southeast Asia, has become a popular pet in the U.S. It can be crossbred with other breeds of pig. The Pot-Bellied pig grows to a smaller stature than other miniatures.

The Sinclair minipig is used for comparative investigations of bone disease by some researchers. Mosekilde and coworkers (1987) report that ovariohysterectomy, combined with moderate calcium restriction, causes loss of bone mass and biomechanical competence, mimicking changes in menopausal women. The same group also reported that fluoride-induced changes were similar in minipigs and humans.

Although expense is a major consideration when miniature swine are selected as an experimental model, there are several vendors of minipigs. Worldwide availability appears to be close at hand. Given the size constraints of farm pigs it may be that minipigs will become a more cost effective alternative in the near future.

Like the rat the pig has a single pair of parathyroids. Porcine calcitonin has been sequenced, with 14 of the 32 amino acids homologous to human calcitonin.

Unlike the sheep, the pig is a true omnivore; in fact, the pig is one of the best models of human gastrointestinal function. Dietarily, pregnant sows require 13.5 g of calcium and 10.8 g phosphorus per day, with a Ca:P ratio of 1–1.5:1 (Capen and Rosol 1989). Younger pigs (10–20 kg) require half as much of these minerals. Nutritional studies are ongoing, perhaps influenced by the surge in the use of the pig as a model for human gastrointestinal disease.

8.7 Nonhuman Primates

That the distinct advantages of using nonhuman primates are constrained by high costs was discussed earlier. The greatest benefit is the similarity of bone mechanics and estrous cycles and menopause to human females. Additionally, their dietary requirements are more like humans than any other domestic species.

A few different primate models of bone loss have been recently published on Old World monkeys. Thompson et al. (1992) studied the effects of a bisphosphonate upon the metabolism in ovariectomized baboons (*Papio* spp.). It was found that the ovariectomized animals experienced increased bone turnover and bone loss of the lumbar spine. Alendronate maintained all measured parameters at control, nonovariectomized levels. It was also reported that male baboons have a finer, more closely woven trabecular network, lower levels of formation, and more extensive erosion than humans (Schnitzler et al. 1993). However, the baboons were captured from the wild and kept in confinement prior to bone biopsy and so the effects of confinement could not be measured.

Mann et al. (1990) administered GNRH by implantable osmotic minipump for 10 months to 13–16-year-old female rhesus macaques *(Macaca mulatta)*. All three animals experienced falling serum progesterone and estradiol levels. This correlated with a gradual reduction in bone mineral density of the caudal vertebrae and humerus, becoming significant after 9 months of treatment. A distinct advantage of this model is the ability to remove the GNRH stimulus and to evaluate the return of menstrual cycles and the recovery of bone mineral density. The effects of estrogen and/or progesterone replacement therapy upon ovariectomized cynomolgus macaques *(Macaca fascicularis)*, all of which were part of a larger study of diet-induced coronary artery disease, has also been examined. Again, bone mineral density was lower in the estrogen-deficient animals, and trabecular plate number was lower while trabecular plate separation rose (Jayo et al. 1990). In the cortical bone of 14 adult cynomolgus macaques (mean age 9.5) years) studied for 2 years postovariectomy there were no significant differences in structural parameters, compared with intact females. However, there were significant changes in intracortical bone, evidenced by increased percent osteonal area, osteonal density, mean osteon area, mean canal area, percent porosity, mean wall width, and mean osteoid area in the femora of the ovariectomized monkeys (Lundon and Grynpas 1993).

Although primates are more similar to humans in reproductive physiology than other species, this depends in part upon their housing. Indoor housing may prevent the anovulatory periods noted in the summer with some animals. Spontaneous osteomalacia has been observed in New World primates housed indoors. This may be the result of inadequate dietary vitamin D_3 intake (Capen and Rosol 1989).

Although there is a considerable amount of general knowledge of nutrient requirments of Old World primates, details of mineral requirements are somewhat lacking. NRC guidelines indicate that growing rhesus macaques require 0.15 g of calcium/kg body weight and 25 IU vitamin D in their diet (Richter et al. 1984). Semipurified diets are available to meet these requirements. Serum levels of calcium and phosphorus have been reported for several primate species, including baboons (Ca = 8.4 ± 1.5 mg/dl, P = 7.0 ± 1.5 mg/dl) and macaques (Ca = 9.7 ± 1.6 mg/dl, P = 6.2 ± 2.09 mg/dl (Loeb and Quimby 1989).

8.8 Conclusion

Choosing the most suitable species is only one concern in the selection of an animal model. Factors such as the age of experimental animals and their average life span are also important considerations, aside even from the costs involved. Factors to take into account also include published normal histomorphometric values, reproductive physiology, dietary similarities, and gastrointestinal absorption.

It appears that in the future the commonly used rat will be replaced more often by other animal species to meet research needs. However, the rat still has its place. A logical approach may be to use the rat for preliminary screening of new pharmacologic agents or therapeutic modalities, followed by verification in other species before undergoing clinical trials in human patients.

The dog which has been used to study estrogen deficiency-induced bone loss for several years has the advantage of being well characterized and easy for most facilities to maintain. It is also an animal species that is large enough to provide multiple bone biopsies. However, these biopsies must be collected carefully and evaluated using appropriate histomorphometric techniques.

It is highly likely that in the future more investigators may turn to sheep or swine as potential animal models, depending on availability. Finally, a few investigators have examined nonhuman primates as models. But they will, for the most part, continue to be impractical alternatives, given the restrictions to their use and high costs.

It should be noted that a number of the studies cited in this paper used dual photon absorptiometry rather than traditional histomorphometry as the major experimental method. This may have a dramatic effect in broadening the alternatives available with animal models since this research tool should increase the number of longitudinal studies that can be performed noninvasively in all species.

References

Capen CC, Rosol TJ (1989) Calcium regulating hormones and diseases of abnormal mineral (calcium, phosphorus, magnesium) metabolism. In: Kaneko JJ (ed) Clinical biochemistry of domestic animals. Academic Press, San Diego, pp 678–752

Chow JWM, Badve S, Chambers TJ (1993) A comparison of microanatomic basis for coupling between bone formation and bone resorption in man and the rat. Bone 14:355–360

Erben RG, Weiser H, Sinowatz F, Rambeck WA, Zucker H (1992) Vitamin D metabolites prevent vertebral osteopenia in ovariectomized rats. Calcif Tissue Int 50(3):228–236

Faugere MC, Malluche HH, Okamoto S, DeLuca HF (1982) Protective effect of 1 25(OH)₂ vitamin D against bone loss in oophorectomized rats. Clin Res 31:741A

Faugere MC, Okamoto S, DeLuca HF, Malluche HH (1986) Calcitriol corrects
    bone loss induced by oophorectomy in rats. Am J Physiol 250:E35–E38
Faugere MC, Friedler RM, Fanti P, Malluche HH (1990) Bone changes occur-
    ring early after cessation of ovarian function in Beagle dogs: a histomor-
    phometric study employing sequential biopsies. J Bone Miner Res
    5(3):263–272
Fraser CM (1986) The Merck veterinary manual. Merck, Rahway, NJ
Ibbotson KJ, Orcutt CM, D'Souza SM, Paddock CL, Arthur JA, Jankowsky
    ML, Boyce RW (1992) Contrasting effects of parathyroid hormone and in-
    sulin-like growth factor I in an aged ovariectomized rat model of post-
    menopausal osteoporosis. J Bone Miner Res 7(4) :425–432
Jayo MJ, Weaver DS, Adams MR, Rankin SE (1990) Effects on bone of surgi-
    cal menopause and estrogen therapy with or without progesterone replace-
    ment in cynomolgus monkeys. Am J Obstet Gyn 163 (2) :614–618
Kalu DN (1991) The ovariectomized rat model of postmenopausal bone loss.
    Bone Miner 15:175–192
Kalu DN, Liu CC, Hardin RR, Hollis BW (1989) The aged rat model of
    ovarian hormone deficiency bone loss. Endocrinology 124:7–16
Kimmel DB, Wronski TJ (1990) Nendestructive measurement of bone mineral
    in femurs from ovareictomized rats. Calcif Tissue Int 46:101–110
Kohn DF, Barthold SW (1984) Biology and diseases of rats. In: Fox JG,
    Cohen BJ, Loew FM (eds) Laboratory animal medicine. Academic Press,
    Orlando, pp 91–122
Loeb WF, Quimby FW (1989) The clinical chemistry of laboratory animals.
    Pergamon, New York
Lopez-Hilker S, Galceran T, Chan UL, Rapp N, Martin KJ (1986) Hypocalce-
    mia may not be essential for development of secondary hyperparathyroid-
    ism due to chronic renal failure. J Clin Invest 78:1097–1102
Lundon K, Grynpas M (1993) The longterm effect of ovariectomy on the
    quality and quantity of cortical bone in the young cynomolgus monkey: a
    comparison of density fractionation and histomorphometric techniques.
    Bone 14:389–395
Malluche HH, Faugere MC (1990) Role of calcitriol in the management of os-
    teoporosis. Metabolism 39 [Suppl 1]:24–26
Malluche HH, Faugere M, Rush M, Friedler R (1986) Osteoblastic insuffi-
    ciency is responsible for maintenance of osteopenia after loss of ovarian
    function in experimental beagle dogs. Endocrinology 119:2649–2654
Malluche HH, Faugere MC, Friedler R, Fanti P (1988) 1,25 Dihydroxyvitamin
    D3 corrects bone loss but suppresses bone remodeling in ovariohysterec-
    tomized beagle dogs. Endocrinology 122:1998–2005

Mann DR, Gould KG, Collins DC (1990) A potential primate model for bone loss resulting from medical oophorectomy or menopause. J Clin Endocrinol Metab 71(1):105–110

Mavrogenis AP, Robison OW (1976) Factors affecting puberty in swine. J Anim Sci 42(5):1251–1255

Monier-Faugere MC, Friedler RM, Bauss F, Malluche HH (1992) A new bisphosphonate BM21.0955 prevents bone loss occurring after cessation of ovarian function in experimental dogs. J Bone Miner Res 7 [Suppl 1]:PS277

Mosekilde L, Kragstrup J, Richards A (1987) Compressive strength, ash weight and volume of vertebral trabecular bone in experimental fluorosis in pigs. Calcif Tiss Int 40:318–322

Nakamura T, Nagai Y, Yamato Suzuki K, Orimo H (1992) Regulation of bone turnover and prevention of bone atrophy in ovariectomized beagle dogs by the administration of 24R 25(OH)2D3. Calcif Tissue Int 50(3):221–227

Owen RA, Melton LJ, Riggs BL, Gallagher JC (1980) The national cost of acute care of hip fractures associated with osteoporosis. Clin Orthop 150:172–176

Park RD, Turner AS, Trotter GW, Aberman HM (1992) Effects of age and ovariectomy on trabecular bone patternof the proximal femur in sheep. Bone 13:A27

Pastoureau P, Arlot ME, Caulin F, Barlet JP, Meunier PJ, Delmas PD (1989) Effects of oophorectomy in biochemical and histological indices of bone turnover in ewes. J Bone Miner Res 14 [Suppl 1]:S237

Pond WG, Houpt KA (1978) The biology of the pig. Cornell University Press, Ithaca, pp 129–180

Richter CB, Lehner NDM, Henrickson RV (1984) Promates. In: Fox JG, Cohen BJ, Loew FM (eds) Laboratory animal medicine. Academic, Orlando, pp 297-383

Rodgers JB, Sherwood LC, Fink BF, Sadove RC (1993) Estrus detection in miniature swine using vaginal cytology. Lab Anim Sci 43(6) (in press)

Schnitzler CM, Ripamonti U, Mesuita JM (1993) Bone histomorphometry in baboons in captivity. Bone (In Press)

Shen V, Dempster DW, Birchman R, Mellish RW, Church E, Kohn D, Lindsay R (1992) Lack of changes in histomorphometric bone mass and biochemical parameters in ovariohysterectomized dogs. Bone 13(4):311–316

Thompson DD, Seedor JG, Quartuccio H, Solomon H, Fioravanti C, Davidson J, Klein H, Jackson R, Clair J, Frankenfield D et al. (1992) The bisphophonate, alendronate, prevents bone loss in ovariectomized baboons. J Bone Miner Res 7: 951–960

Turner CH (1991) Homeostatic control of bone structure: an application of feedback theory. Bone 12(3):203–217

Wessler S (1976) Introduction: what is a model? In: Animal models of thrombosis and hemorrhagic diseases: report of a workshop organized by the Institute of Laboratory Animal Resources (ILAR) Committee on Animal Models for Thrombosis and Hemorrhagic Diseases. US Dept. of Health Education and Welfare, Washington, pp xi–xvi

Wronski TJ, Schenck PA, Cintron M, Walsh CC (1987) Effect of body weight on osteopenia in ovariectomized rats. Calcif Tissue Int 10: 155–160

9 The Rat as an Animal Model of Postmenopausal Bone Loss

Dike N. Kalu

9.1 Introduction

With the continuing demographic shift in population toward a more aged society, age-related diseases such as osteoporosis have emerged as major public health problems. Osteoporosis is the end point of the age-related bone loss that occurs in all human populations. It is characterized by low bone mass and increased susceptibility to fracture from minor trauma. In the United States alone osteoporosis afflicts 15–20 million people and results in fractures in 1.3 million people 45 years and older [1]. The societal cost of managing osteoporosis has been estimated at over 7 billion dollars per year [2]. The price tag will continue to increase if effective measures are not developed to identify those at risk and prophylactic or therapeutic measures instituted.

Osteoporosis is a heterogeneous disease [3]. By far the most common form of the disease is postmenopausal osteoporosis, also termed type 1 osteoporosis. The major risk factor for type 1 osteoporosis is loss of ovarian hormones at menopause. Although type 1 osteoporosis has been linked to the loss of estrogen, the mechanism by which estrogen deficiency causes osteoporosis remains unclear. Part of the reason for the slow progress in unraveling the etiology of postmenopausal osteoporosis is the lack of generally accepted, convenient animal models for studying bone loss due to ovarian hormone deficiency. One of the best studied of the animal models is the ovariectomized rat [4, 5].

9.2 Aged and Mature Ovariectomized Rat Models

The rat model of postmenopausal bone loss is based on the fact that like humans, female rats experience accelerated loss of bone when they lose their sex hormones. The hormone deficiency is usually induced by surgical ovariectomy, although rats can also be chemically castrated by treatment with luteinizing hormone-releasing hormone (LHRH) agonists. Two types of ovariectomized rat models can be distinguished based on the age of the rats, namely, the "aged rat model" and the "mature rat model" [4]. The aged rat model is based on animals aged 12 months and older. Aged rats are used to model human postmenopausal bone loss because human bone loss is age related and begins long after the attainment of skeletal maturity. It is only appropriate that models of age-related bone loss be based on rats whose skeleton is not growing rapidly. It should, however, be pointed out that 12-month-old rats are not really old relative to the lifespan of rats, which is about 3–4 years depending on the strain. The aged rat model is dubbed "aged" only to denote that the skeletal characteristics of 12-month-old rats have stabilized at levels close to those of old rats [4]. Aged rats are expensive, difficult to obtain, and take too long to lose significant amounts of bone following ovariectomy. In contrast, "mature" rats, aged about 3 months, do not have these shortcomings. Consequently, they are widely used to study ovariectomy-induced bone loss. The term "mature" rat model is used to denote that the rats, though young, are reproductively competent and, therefore, respond appropriately to sex hormone deficiency following ovariectomy [4].

9.3 The Rat Model and Bone Remodeling

Although many investigators had demonstrated that rats lose bone following ovariectomy, the ovariectomized rat was not immediately accepted as an appropriate model of postmenopausal bone loss. The apparent reasons for this are many [4], but the most important relates to the question of bone remodeling in rats. The 1960s and 1970s witnessed the realization that bone remodeling is based on basic multicellular units (BMU) and that human osteoporosis is a disease of impaired BMU-based bone remodeling. It was argued that rats lack useful amounts of BMUs and, therefore, cannot provide good models of human bone disease or its treatment [6]. Recent reviews on the ovariectomized rat bone loss model indicate that this latter view is no longer tenable [4, 5]. Furthermore, Frost and Jee [6] have also emphasized in a recent review that the same mechanisms control BMU-based remodeling in rats and humans and that as a consequence the rat can provide a useful model for human bone loss. "The opinion that it would not stemmed from incomplete knowledge, and it errs" [6].

9.4 Similarities Between Postmenopausal Bone Loss and the Ovariectomized Rat Bone Loss Model

There are extensive similarities in the characteristics of bone loss in postmenopausal women and the ovariectomized rat model [4–6] (Table 1). Briefly, in both species, ovarian hormone deficiency results in low bone mass due to an increase in the rate of bone turnover with resorption exceeding formation; the increased rate of bone turnover is evident in increased biochemical indices; the rate of cancellous bone loss exceeds that of cortical bone loss; in rats the increased cancellous bone loss, which as in humans involves BMUs, is found mainly in the metaphysis of the long bones rather than in the vertebrae, which loses cancellous bone less rapidly in rats. Every drug that has been found to

Table 1. Comparison of postmenopausal bone loss with the ovariectomized rat bone loss model

Parameters	Postmenopausal bone loss	Ovariectomized rat bone loss model
Ovarian hormone status	↓	↓
Bone mass	↓	↓
Rate of bone turnover	↑	↑
Resorption > Formation	Yes	Yes
Biochemical indices of bone turnover	↑	↑
Cancellous bone loss > cortical bone loss	Yes	Yes
Bone loss involves BMUs	Yes	Yes
Bones respond similarly to identical therapies	Yes	Yes
Intestinal calcium absorption	↓	↓
Calcium-regulating hormones	↔/↓	↔/↓
Primarily involvement of calcium-regulating hormones	Unlikely	Unlikely
Involvement of hematopoietic system	Likely	Likely
Involvement of cytokines	Likely	Likely

BMU, basic metabolic units; ↓, decreased; ↑, increased; ↔, unchanged.

modulate bone loss in postmenopausal women also retards bone loss in the rat model. The intestinal calcium malabsorption that occurs in postmenopausal osteoporosis also occurs in the aged rat model in both humans and rats the calciuim-regulating hormones appear not to play the primary role in the etiology of the bone loss. Rather, recent findings indicate that in both species it is most likely that bone loss due to ovarian hormone deficiency involves perturbations of the hematopoietic system and increased secretion of cytokines that stimulate bone resorption.

9.5 Applications of the Rat Model

There are, at least, two main reasons for seeking a rat model of postmenopausal bone loss. Because the cause of the bone loss is uncertain, the first is to employ the model to seek the underlying mechanisms for postmenopausal bone loss. An understanding of the etiology of the bone loss will likely result in the design of rational therapies or effective preventive interventions for postmenopausal osteoporosis. Since a large body of patients already have osteoporosis, with the number growing as the population of aged people increases, the second use of the rat model is for evaluating potential therapies for postmenopausal osteoporosis based on our current understanding of bone biology and their mechanisms of action. Investigators from many laboratories are using the rat model to address the above two issues. The account that follows focuses mainly on some of the studies which have been done in my laboratory in these areas, as the contributions of other investigators are presented in other chapters in this publication.

9.6 Testing the Altered Calcium Homeostatic Control Hypotheses

When we became interested in a model for postmenopausal bone loss, the prevailing hypotheses for postmenopausal osteoporosis centered around the linkage of the bone loss to alterations in the levels or actions of the calcium-regulating hormones [7–15]. According to one view that emerged from the work of many investigators [7–13], estrogen normally suppresses bone resorption; consequently, loss of ovarian hormones as occurs following menopause results, sequentially, in increased calcium release from bone, suppression of parathyroid hormone (PTH) secretion by calcium, decreased $1,25(OH)_2$ vitamin D levels as a result of the inhibition of the activity of kidney 25-(OH)D-1-α-hydroxylase due to low PTH, and decreased calcium absorption, leading to bone loss. In another scenario estrogen is envisioned to have a tropic effect on calcitonin synthesis and secretion [14]. Since calcitonin is a potent inhibitor of bone resorption, a fall in its level following menopause will leave PTH-induced bone resorption to proceed unopposed, resulting in bone loss. Others speculate that the increase that occurs with aging in PTH levels per se will predispose to age-related bone loss [15]. Because these hypotheses, for the most part, have in common the linkage of ovarian hormone deficiency bone loss with alterations in the levels or activities of the calcium-regulating hormones, we have grouped them under the heading, "altered calcium homeostatic control hypotheses" for the pathogenesis of bone loss due to ovarian hormone deficiency. To test these hypotheses, we evaluated the effects of ovariectomy on bone and the calcium-regulating hormones in 12-month-old rats fed normal or low dietary calcium [16]. After 6 months, ovariectomy and low calcium diet independently decreased the density of the ilium, the femur, and the fourth lumbar vertebra as well as the calcium content of the latter two. Low calcium diet increased morphometric indices of bone formation and bone resorption, as did ovariectomy; resorption must have exceeded formation to account for the net loss of bone. The effects of ovariectomy and low calcium diet on bone were additive. Ovariectomy had no effect on serum levels of PTH and vitamin D metabolites oron serum calcitonin in rats fed a low calcium diet, while the latter caused a significant increase in serum 1,25-dihydroxyvitamin D levels. In both dietary regimens ovariectomy resulted in about 30% decrease in intestinal calcium absorption. The findings indicate that while ovariectomy induces bone loss in aged rats, it does not mediate this effect by altering the levels of the calcium-regulating hormones, PTH, calcitonin, or the vitamin D metabolites [16]. However, serum calcitonin was decreased by ovariectomy in animals fed normal calcium diet. To ensure that the bone loss observed in ovariectomized animals fed normal calcium diet was not due to the observed low calcitonin levels, we carried out additional experiments in animals bearing parathyroid autotransplants and thyroidectomized to remove their endogenous source of calcitonin [17]. Bone loss still occurred following ovariectomy, in spite of calcitonin deficiency in both control and ovariectomized animals, confirming that the calcium-regulating hormones are not the primary etiologic factors in bone loss due to ovarian hormone deficiency.

In a recent study from Japan, women who were oophorectomized for therapeutic reasons had decreased total body and spinal bone mineral density; however, their serum levels of PTH, calcitonin and $1,25(OH)_2$ vitamin D were not significantly different from those of age-matched controls [18]. In another study of sex differences in bone loss in the second metacarpal, both men and women lost cortical bone with age, but the rate of bone loss was greater in women at the time of menopause [19]. In contrast, there were no sex differences in the pattern of age-related changes in the calcium-regulating hormones. The investigators concluded that, "it is unlikely that the increased bone loss noted in females can be explained by alterations in vitamin D status and the circulating levels of the bone mineral regulating hormones" [19]. It is clear that both human studies and studies based on the rat model agree that the altered calcium homeostatic control hypotheses are inadequate to explain the underlying basis for the bone loss due to ovarian hormone deficiency.

9.7 Estrogen Receptor in the Intestine

Postmenopausal osteoporosis is often associated with decreased calcium absorption that is believed to contribute to or aggravate the bone loss. The exact cause of the calcium malabsorption is not clear but has been attributed to decreased serum $1,25(OH)_2$ vitamin D and to intestinal resistance to its action. In the course of characterizing the aged rat model, we observed that the bone loss due to ovariectomy was associated with decreased intestinal absorption of calcium [16]. Since in the study, $1,25(OH)_2$ vitamin D levels were not altered, it is conceivable that the decreased calcium absorption resulted directly from low estrogen levels. We, therefore, hypothesized that estrogen acts directly on intestinal cells to enhance calcium absorption through estrogen receptor mediated mechanisms. The studies we carried out to test this hypothesis demonstrated that the intestinal mucosal cells contain estrogen receptor immunoreactivity, express the mRNA for estrogen receptor, and respond directly to 17β-estradiol with enhanced calcium transport that is suppressed by gene transcription and protein synthesis inhibitors [20]. These findings are novel and suggest that estrogen has a physiological role in the regulation of intestinal calcium absorption, and that its deficiency in postmenopausal osteoporosis and following therapeutic ovariectomy may result directly in the calcium malabsorption that is believed to be an important factor in the bone loss that occurs in these conditions.

9.8 Bone Loss and the Hematopoietic System

The "altered calcium homeostatic control hypotheses" was found to be inadequate in explaining the underlying basis for the bone loss due to ovarian hormone deficiency. Consequently, alternative hypotheses are being sought based on recent advances in bone biology. These recent advances include evidence that (a) bone contains a plethora of local factors that are likely involved in the regulation of bone metabolism [21]; (b) immune and bone marrow cells synthesize and secrete cytokines that may have autocrine and paracrine influence on bone metabolism [22, 23]; (c) bone marrow contains hematopoietic precursors of osteoclasts [24–26]; (d) ovariectomy increases the proliferation of these osteoclast-like progenitors [27–29]; (e) mononuclear bone marrow cells [30] and peripheral blood monocytes [31] probably have the capacity to resorb bone; (f) alterations in peripheral T lymphocyte subsets are associated with decreased bone mass in postmenopausal women [32]; (g) ovarian hormone deficiency results in enhanced monocyte cytokine production and increased bone turnover that are suppressed by estradiol [29, 33–35].

The above findings support the association of altered bone metabolism with elements of the hematopoietic system and their cytokines and form the basis for altered hematopoiesis–local factor hypothesis for the pathogenesis of bone loss due to ovarian hormone deficiency. According to the hypothesis, ovarian hormones have a pleiotropic effect on the hematopoietic system; consequently, deficiency of these hormones, as occurs following menopause or ovariectomy, results in alterations in the populations and/or activities of bone-related blood cells, especially in bone marrow; these alterations by themselves or through their interactions with bone cells change the balance of bone turnover in favor of bone loss. Components of this hypothesis are the subject of recent reports [27, 29, 36, 37].

To investigate the relationship of the hematopoietic system to the loss of bone due to ovarian hormone deficiency, we examined the effects of ovariectomy and estrogen administration on the thymus, spleen, and bone marrow and on the proliferation of marrow progenitors of osteoclasts [37, 38]. Ovariectomy, in the mature rat model, resulted in decreased cancellous bone volume, increased trabecular osteoblast and osteoclast numbers, and increased serum alkaline phosphatase levels that were prevented by 17β-estradiol treatment. Thymus weight, spleen weight, thymus and spleen lymphocytes, and bone marrow monocytes and lymphocytes also increased significantly following ovariectomy, and the increases were suppressed by 17β-estradiol. Our findings indicate that ovariectomy-induced bone loss in the rat is accompanied by marked changes in the hematopoietic system, and that these changes are modulated by estrogen administration. These findings in ovariectomized animals are of note because monocytes secrete cytokines such as interleukin-1 (IL-1) which stimulates bone resorption. The occurrence of increased monocyte population in the microenvironment of bone marrow is ideally suited for enhanced direct interaction of IL-1 with neighboring cells and with other local factors to alter the rate of bone metabolism.

In order to further explore the contribution of the hematopoietic system to the excessive osteoclastic bone resorption that results in osteoporosis in postmenopausal women, we cultured bone marrow cells from ovariectomized and sham-operated mice for 8 days [27]. The cells gave rise in culture to tartrate-resistant acid phosphatase-positive multinucleate cells. The formation of these osteoclast-like cells was enhanced by PTH and $1,25(OH)_2$ vitamin D_3, with the latter being more effective. Cultures of cells from ovariectomized animals formed significantly more tartrate-resistant acid phosphatase-positive multinucleate cells than those from sham-operated controls. These findings support our hypothesis that ovarian hormone deficiency promotes the expansion of a pool of marrow-derived hematopoietic progenitor cells that differentiate into bone-resorbing osteoclasts under the influence of osteotropic hormones. These studies, which were first carried out in mice, have been confirmed in the ovariectomized rat bone loss model [38, 39] and extended by others as described by Manolagas in Chap. 6 (this volume). The above findings indicate that the nature of the involvement of the hematopoietic system in the bone loss due to ovarian hormone deficiency merits continued exploration.

9.9 The Rat Model and Transcription of Cancellous Bone Proteins

Ovariectomy-induced bone loss is caused by increased bone turnover in which resorption exceeds bone formation. The marrow-related studies described above will help to explain the increased osteoclastic bone resorption aspect of the increased bone turnover. In contrast, little is known about the osteoblast component, beyond some report that os-

teogenic precursors in marrow are also increased following ovariectomy [40]. To address this deficit, we have established a technique for
examining the gene expression of osteoblast-related proteins in bone.
We can now isolate RNA from cancellous bone, a very sensitive target
of ovarian hormone deficiency in both postmenopausal women and the
ovariectomized rat bone loss model. We have used the technique to
examine the effects of ovariectomy and 17β-estradiol on osteoblast related mRNA pools of cancellous bone proteins of the distal metaphysis
of the femur [41]. Ovariectomy resulted in a significant increase in the
mRNAs of type 1 collagen, osteocalcin and c-myc. In addition, ovariectomy caused the expected decrease in cancellous bone in the proximal tibia, and increased osteoclast and osteoblast numbers. The increases were suppressed in animals that received 17β-estradiol
injections. These findings suggest that the lack of ovarian hormones
shortly after ovariectomy upregulates, and estrogen administration
downregulates the expression of important cancellous bone matrix proteins as well as the protooncogen, c-myc. Similar observations have
been made by Turner et al. [42]. How the interpretation of the bone
mRNA data will be influenced by the finding that ovariectomy and estrogen therapy also altered bone cell numbers is presently unknown.

9.10 The Rat Model and Potential Therapies for Postmenopausal Bone Loss

The presentation, so far, has dealt with the application of the rat model
to probe the underlying mechanisms of bone loss due to ovarian hormone deficiency. The second application of the ovariectomized rat
bone loss model is to use the model for evaluating the efficacy of
potential therapies for postmenopausal osteoporosis and their mechanisms of action. Therapies for osteoporosis can be divided into two
groups: those that inhibit bone resorption and those that stimulate bone
formation. Drugs that inhibit bone resorption are particularly useful for
treating bone loss characterized by increased bone turnover. The best
studied of such drugs are estrogenic based drugs, bisphosphonates and
calcitonin. After bone loss has occurred and bone turnover decreased,
drugs that stimulate bone formation are required to rebuild the bone

that has been lost. There is a great deal of interest in seeking for such drugs since none is currently in general use clinically.

Insulin like growth factor-I (IGF-I) is of great interest as a potential therapy for established bone loss since it is believed to mediate the somatic and bone anabolic actions of growth hormone. We have carried out three studies to evaluate the actions of IGF-I in the ovariectomized rat bone loss model. In the first two studies, IGF-I administration was started on the day of ovariectomy. In the third study, IGF-I administration, with and without its binding protein IGFBP-3, was initiated 30 days after ovariectomy, when bone loss due to ovarian hormone deficiency had already occurred. In the first, but not the second, study, IGF-I had a modest effect in preventing bone loss due to ovariectomy [43]. In the third study, IGF-I and IGFBP-3, alone or in combination, were not effective in rebuilding bone after loss due to ovariectomy had already occurred [44]. The studies were carried out with the mature rat model. Similar negative findings were reported on the action of IGF-I on bone loss in aged ovariectomized rats [45]. These therapeutic failures are surprising in view of the well-known bone anabolic actions of IGF-I, which is believed to mediate the effects of growth hormone on bone. However, some success with IGF-I was reported in an abstract of studies carried out on immature ovariectomized rats [46].

In view of the largely negative findings with IGF-I, we reasoned that, regarding the treatment of ovariectomy-induced bone loss, stimulating the production of IGF-I locally in bone may be superior to just increasing its systemic level in blood by parenteral administration as in the above studies. For this reason, we decided to examine the therapeutic potential of growth hormone (GH) in reversing ovariectomy-induced bone loss since GH is known to cause a local increase in IGF-I synthesis in bone. Furthermore, GH may have other salutary effects on bone anabolism besides stimulating IGF-I production. Recombinant human GH administration was initiated 30 days postovariectomy, after bone loss had already occurred in the mature rat model. Two dose levels of GH, 2 mg and 8 mg/kg body weight, were given daily, subcutaneously, for 30 days; the animals were killed and their bones examined by histomorphometry. The effects of GH in rebuilding bone after loss due to ovariectomy were dramatic. The low dose GH (2 mg) returned trabecular bone volume toward the level for sham-operated control animals; the high dose GH (8 mg) completely returned the bone mass of the ovariectomized animals to a level similar to those of the sham controls. These findings indicate that the potential of GH for stimulating bone formation in ovarian hormone deficiency states merits continued exploration.

References

1. Anonymous (1984) Osteoporosis: consensus conference. JAMA 252:799–802
2. Holbrook TL, Grazier K, Kelsey JL, Stauffer RN (1984) The frequency of occurrence, impact and cost of selected musculoskeletal conditions in the United States. American Academy of Orthopedic Surgeons, Chicago
3. Melton LJ III, Riggs BL (1988) Clinical spectrum. In: Riggs BL, Melton LJ III (eds) Osteoporosis: etiology, diagnosis and management. Raven, New York, pp 155–179
4. Kalu DN (1991) The ovariectomized rat model of postmenopausal bone loss. Bone Miner 15:175–192
5. Wronski TJ, Yen CF (1991) The ovariectomized rat as an animal model for postmenopausal bone loss. Cells Materials [Suppl] 1:69–74
6. Frost HM, Jee WSS (1992) On the rat model of human osteopenias and osteoporoses. Bone Miner 18:227–236
7. Heaney RP (1965) A unified concept of osteoporosis. Am J Med 39:877–880
8. Young MM, Jasani C, Smith DA, Nordin BEC (1968) Some effects of ethinyl oestradiol on calcium and phosphorus metabolism in osteoporosis. Clin Sci 34:411–417
9. Heaney RP (1969) A unified concept of osteoporosis: a second look. In: Barzel US (ed) Osteoporosis. Grune and Stratton, New York, pp 257–265
10. Heaney RP (1974) Pathophysiology of osteoporosis: implications for treatment. Tex Med 70:37–45
11. Gallagher JC, Riggs BL, Eisman J, Hanstra A, Arnaud SB, DeLuca HF (1979) Intestinal calcium absorption and serum vitamin D metabolites in normal subjects and osteoporotic patients. Effect of age and dietary calcium. J Clin Invest 64:729–736
12. Riggs BL, Melton LJ III (1983) Evidence for two distinct syndromes of involutional osteoporosis. Am J Med 75:899–901
13. Buchanan JR, Cauffman SW, Cireer RB III (1984) Relation of calcium-regulating hormones to the pathogenesis of postmenopausal osteoporosis. In: Christiansen C, Arnand CD, Nordin BEC, Parfitt AM, Peck WA, Riggs BL (eds) Osteoporosis 1. Stifsbogtrykkeri Glostrup, Aalborg, pp 275–280

14. Stevenson JC (1985) Differential effects of aging and menopause on CT secretion. In: Pecile A (ed) Calcitonin 1984. Excerpta Medica, pp 145–152

15. Johnson CC Jr, Epstein S (1982) The endocrinology of osteoporosis. In: Parsons JA (ed) Endocrinology of calcium metabolism. Raven, New York, pp 467–484

16. Kalu DN, Liu CC, Hardin RR, Hollis BW (1989) The aged rat model of ovarian hormone deficiency bone loss. Endocrinology 124:7–16

17. Kalu DN, Hardin RR (1984) Evaluation of the role of calcitonin deficiency in ovariectomy-induced osteopenia. Life Sci 34:2394–2398

18. Ohta H, Makita K, Suda Y, Ikeda T, Masuzawa T, Nozawa S (1992) Influence of oophorectomy on serum levels of sex steroids and bone metabolism and assessment of bone mineral density in lumbar trabecular bone by QCT-C value. J Bone Miner Res 7:659–665

19. Sherman SS, Hollis BW, Tobin JD (1990) Vitamin D status and related parameters in a healthy population: the effects of age, health and season. J Clin Endocrinol Metab 71:405–413

20. Arjmandi BH, Salih MA, Herbert DC, Sims SH, Kalu DN (1993) Evidence for estrogen receptor-linked calcium transport in the intestine. Bone Miner 21:63–74

21. Centrella M, Canalis E (1985) Local regulators of skeletal growth: a perspective. Endocr Rev 6:544–551

22. Horowitz M, Vignery A, Gershon R, Baron R (1984) Thymus-derived lymphocytes and their interactions with macrophages are required for the production of osteoclast-activating factor in the mouse. Proc Natl Acad Sci USA 81:2181–2185

23. Boyce B, Aufdemorte T, Garret J, Yates A, Mundy G (1989) Effects of interleukin 1 on bone turnover in normal mice. Endocrinology 125:1142–1150

24. McDonald B, Takahashi N, McManus L, Holahan J, Mundy G, Roodman G (1987) Formation of multinucleated cells that respond to osteotropic hormones in long term human bone marrow cultures. Endocrinology 120:2326–2333

25. Roodman G, Ibbotson K, MacDonald B, Kuehl T, Mundy G (1985) 1,25-Dihydroxyvitamin D_3 causes formation of multinucleated cells with several osteoclast characteristics in cultures of primate marrow. Proc Natl Acad Sci USA 82:8213–8317

26. Takahashi N, Yamana H, Yoshiki S, Roodman GD, Mundy GR, Jones SJ, Boyde A, Suda T (1988) Osteoclast-like cell formation and its regulation by osteotropic hormones in mouse bone marrow cultures. Endocrinology 122:1373–1382

27. Kalu D (1990) Proliferation of tartrate-resistant acid phosphatase positive multinucleate cells in ovariectomized animals. Proc Soc Exp Biol Med 195:70–74

28. Kalu D, Echon R, Hollis B (1990) Modulation of ovariectomy-related bone loss by parathyroid hormone in rats. Mech Ageing Dev 56:49–62

29. Jilka RL, Hangoc G, Girasol G, Passeri G et al (1992) Increased osteoclast development after estrogen loss: mediation by interleukin-6. Science 257:88–91

30. Hattersley G, Chambers T (1989) Generation of osteoclastic function in mouse bone marrow cultures: multinuclearity and tartrate-resistant acid phosphatase are unreliable markers for osteoclastic differentiation. Endocrinology 124:1689–1696

31. Kahn A, Stewart C, Teitelbaum S (1978) Contact-mediated bone resorption by human monocytes in vitro. Science 199:988–990

32. Imai Y, Tsunenari T, Fukase M, Fujita T (1990) Quantitative bone histomorphometry and circulating T lymphocytes subsets in postmenopausal osteoporosis. Bone Miner Res 5:393–399

33. Pacifici R, Rifas L, McCracken R, Vered L, McMurtry C, Avioli LV, Peck WA (1989) Ovarian steroid treatment blocks a postmenopausal increase in blood monocyte interleukin 1 release. Proc Natl Acad Sci USA 86:2398–2402

34. Pacifici R, Brown C, Rifas L, Avioli L (1990) TNF-A and GM-CSF secretion from human blood monocytes: effects of menopause and estrogen replacement. J Bone Miner Res 5:145–151

35. Pacifici R, Brown C, Puscheck E, Friedrich E, Slatopolsky E, Maggio D, McCracken R, Avioli LV (1991) Effect of surgical menopause and estrogen replacement on cytokine release from human blood mononuclear cells. Proc Natl Acad Sci USA 88:5134–5138

36. Pacifici R (1992) Is there a causal role for IL-1 in postmenopausal bone loss? Calcif Tissue Int 50:295–200

37. Kalu DN, Salerno E, Liu CC, Echon RM, Ray M (1991) The altered hematopoiesis hypothesis for the pathogenesis of ovarian hormone deficiency bone loss. J Bone Miner Res 6 [Suppl 1]:S221

38. Kalu DN, Salerno E, Liu CC, Ferarro F, Arjmandi BH, Salih MA (1990) Ovariectomy-induced bone loss and the hematopoietic system. Bone Miner (in press)

39. Kalu DN, Echon R, Hollis BW (1990) Modulation of ovariectomy-related bone loss by parathyroid hormone in rats. Mech Ageing Dev 56:49–62

40. Egise D, Martin D, Neve P, Vienne A, Verhas M, Schoutens A (1992) Bone blood flow and in vitro proliferation of bone marrow and trabecular bone osteoblast-like cells in ovariectomized rats. Calif Tissue Int 50:336–341

41. Salih MA, Liu CC, Kalu DN (1992) Estrogen modulates the expression of type 1 alpha collagen, osteocalcin and the protooncogene, c-myc, in cancellous bone of ovariectomized rats. J Bone Miner Res 7 [Suppl 1]:S273

42. Turner RT, Colvard DS, Speisberg TC (1990) Estrogen inhibition of periosteal bone formation in rat long bones: downregulation of gene expression for bone matrix proteins. Endocrinology 127:1346–1351

43. Kalu DN, Liu CC, Salerno E, Salih M, Echon R, Ray M, Hollis BW (1991) Insulin like growth factor-1 partially prevents ovariectomy induced bone loss: a comparative study with human parathyroid hormone-(1-38). J Bone Miner Res 6 [Suppl 1]:S221

44. Kalu DN, Liu CC, Arjmandi BH, Salerno E, Salih MA, Hollis BW (1993) Growth hormone but not recombinant human IGF-I reversed bone loss due to ovariectomy in rats. J Bone Miner Res 8 [Suppl 1]:S619

45. Ibbotson KJ, Orcutt CM, D'Souza SM, Paddok CL, Arthur JA, Jankowsky ML, Boyce RW (1992) Contrasting effects of parathyroid hormone and insulin like growth factor-1 in an aged ovariectomized rat model of postmenopausal osteoporosis. J Bone Miner Res 7:425–432

46. Mueller K, Coresi R (1991) Insulin-like growth factor-1 increases trabecular bone in ovariectomized rats. J Bone Min Res 6 [Suppl 1]:S221

10 Histomorphometric Studies of Bone Changes in Ovariectomized Rats and Their Prevention by Estrogen Replacement

Thomas J. Wronski and Chiung-Fen Yen

10.1 Introduction

Use of the ovariectomized (OVX) rat as an animal model for post-menopausal bone loss has increased markedly during the past decade. This report describes osteopenic changes in OVX rats and the bone-protective effect of estrogen replacement, as determined by histomorphometric techniques. The skeletal effects of estrogen depletion and replacement in OVX rats are compared briefly to those in postmenopausal and oophorectomized women.

10.2 Skeletal Effects of Estrogen Depletion

Bone loss is a well-known consequence of ovariectomy in rats (Kalu 1991). The osteopenic changes are much more pronounced in cancel-

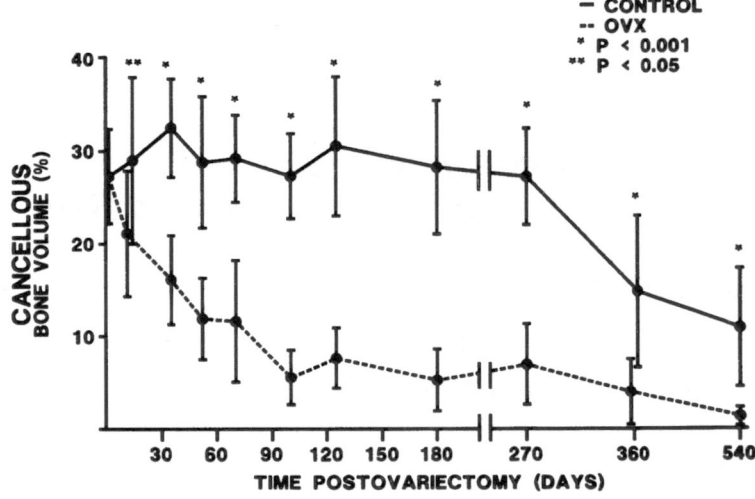

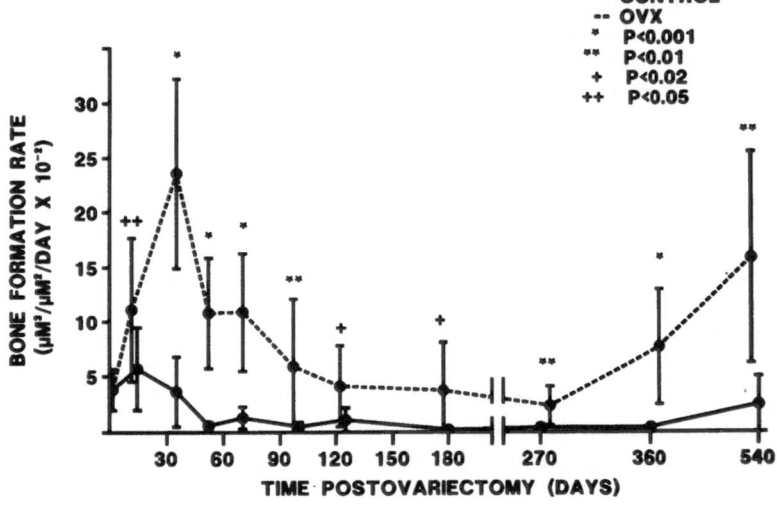

lous bone than in cortical bone. In rats subjected to ovariectomy at 3 months of age, cancellous bone volume in the proximal tibial metaphysis declines from the presurgery level of 25%–30% to the markedly osteopenic level of 5% by 3 months postovariectomy (Fig. 1). This initial, rapid phase of cancellous bone loss is followed by an intermediate period during which tibial bone volume appears to stabilize in OVX rats at approximately 5%. A late, slow phase of cancellous bone loss then begins at 9 months postovariectomy. The proximal tibial metaphysis becomes almost devoid of cancellous bone as the bone volume declines to 1% by 21 months postovariectomy (Wronski et al. 1989a).

The initial, rapid phase of cancellous bone loss in OVX rats is associated with significant increases in osteoclast and osteoblast surfaces. Mineralizing surface, mineral apposition rate, and bone formation rate (tissue level, surface referent) are also increased during the early stages of estrogen depletion in rats (Fig. 1). A transient increase in longitudinal bone growth also occurs in OVX rats, but this skeletal process returns to normal by 1 month postovariectomy. Similarly, the above histomorphometric indices of increased bone turnover in OVX rats decline toward control levels between 3 and 9 months postovariectomy. However, during the late, slow phase of cancellous bone loss, bone turnover again increases in OVX rats to a level well above that of intact, control rats (Fig. 1). Therefore, both the initial rapid phase and late slow phase of cancellous bone loss in the proximal tibia of OVX rats are clearly associated with increased bone turnover (Wronski et al. 1989a). By inference, the increment in bone resorption exceeds the increment in bone formation for net bone loss to occur.

The time course of osteopenic changes in OVX rats has also been characterized in the axial skeleton (Wronski et al. 1989b). Although

---

Fig. 1A,B. Cancellous bone volume (A) and tissue level, surface referent bone formation rate (B) in the proximal tibial metaphysis are plotted as a function of time postovariectomy. Each data point for the control (*solid lines*) and OVX (*broken lines*) groups is the mean ± SD of 10–12 animals, with the exception of the baseline control group at day 0 ($n = 8$) and the control group at 540 days ($n = 7$). Levels of significance were determined with the two-tailed Student's t test. Note the close temporal association between the early and late phases of cancellous bone loss in OVX rats and increased bone formation rate, an index of bone turnover (From Wronski et al. 1989a)

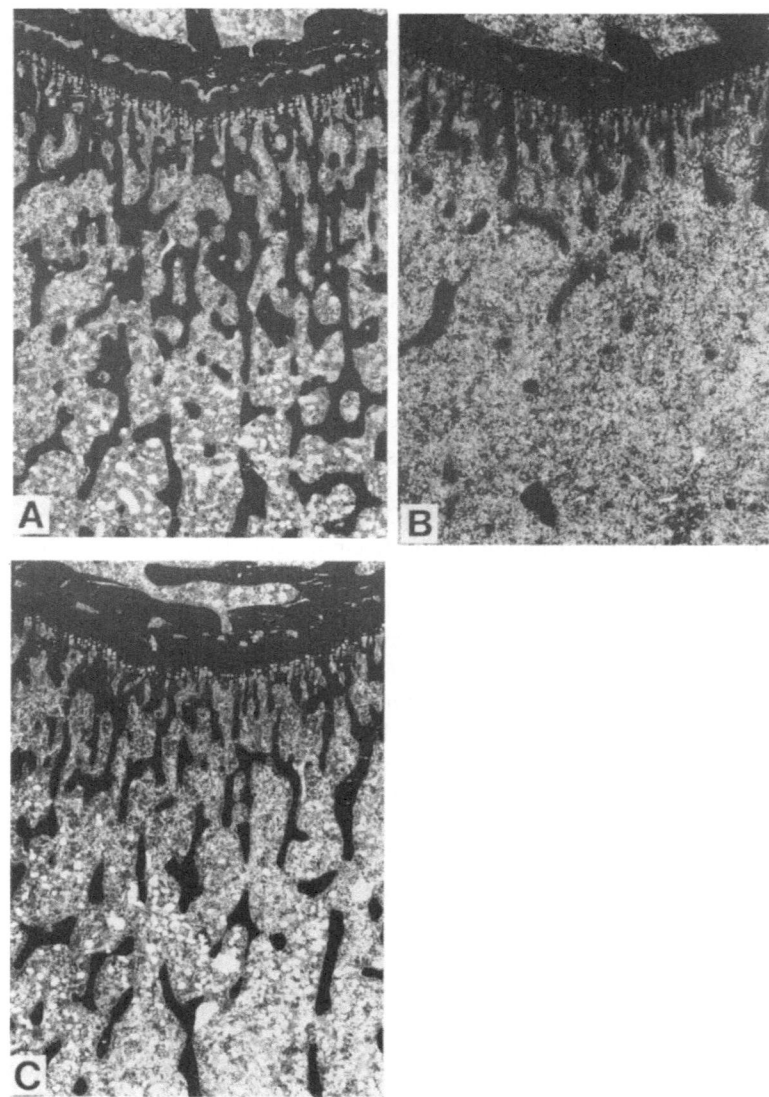

cancellous bone loss and increased bone turnover occur in the lumbar vertebral bodies of OVX rats, the osteopenia is slower to develop and is not as pronounced as in the proximal tibial metaphysis. For example, loss of 50% of cancellous bone does not occur until 9 months postovariectomy in the lumbar vertebra, whereas bone loss of the same magnitude occurs by the end of the first month postovariectomy in the proximal tibia. Despite this temporal difference, the axial and appendicular skeletons of rats respond similarly to estrogen depletion by the development of cancellous osteopenia and increased bone turnover.

Similar to OVX rats, osteopenic changes in early postmenopausal and oophorectomized women occur primarily in cancellous bone (Genant et al. 1982; Nilas et al. 1984). Estrogen-depleted women also exhibit an initial rapid phase of bone loss after which the rate of bone loss slows substantially (Horsman et al. 1977; Elders et al. 1988: Nilas and Christiansen 1988). For evaluation of bone turnover, histomorphometric data are not yet available from women specifically during the early stages of estrogen depletion.

Nevertheless, abundant biochemical data clearly indicate an increase in bone turnover in women during this time (Heaney et al. 1978; Johnston et al. 1985; Riis et al. 1986; Stepan et al. 1987; Kelly et al. 1989). Several studies suggest that bone turnover peaks during the first 5–10 years of estrogen depletion in women, followed by a decline toward premenopausal or preoophorectomy levels (Stepan et al. 1987; Kelly et al. 1989). Therefore, the early skeletal effects of estrogen depletion in rats and women are qualitatively very similar.

Fig. 2A–C. Proximal tibial metaphyses from control (**A**) and OVX (**B**) rats treated with vehicle alone, and from an OVX rat treated with estrogen (**C**). Treatments were initiated on the day after surgery and continued for a 35-day period. Estrogen treatment consisted of subcutaneous injections of 17β-estradiol 5 days/week at a dose of 10 µg/kg body weight. Note the decreased amount of darkly stained cancellous bone spicules indicative of osteopenia in the vehicle-treated OVX rat and the preservation of these spicules in the estrogen-treated OVX rat. (From Wronski et al. 1989c)

10.3 Skeletal Effects of Estrogen Replacement

Treatment of OVX rats with estrogen soon after surgery prevents the development of cancellous osteopenia (Wronski et al. 1988; Kalu et al. 1991; Turner et al. 1993). The bone protective effect of estrogen (Fig. 2) has been observed in both the axial and appendicular skeletons of OVX rats during studies lasting as long as 1 year (Wronski et al. 1991). OVX rats treated with estrogen exhibit decreased longitudinal bone growth and suppressed histomorphometric indices of bone formation and resorption (i.e., decreased bone turnover). In short, estrogen replacement normalizes cancellous bone mass and bone turnover in OVX rats to the level of intact, control rats. It is important to note that cancellous bone is lost rapidly, and that bone turnover increases markedly soon after withdrawal of estrogen treatment in OVX rats (Wronski et al. 1993b). In contrast to its impressive bone prophylactic effect, estrogen failed to restore lost cancellous bone in osteopenic OVX rats (Wronski et al. 1993a).

Prevention of bone loss in postmenopausal women by estrogen replacement is well-documented (Recker et al. 1977; Lindsay et al. 1978; Christiansen et al. 1982; Thomsen et al. 1986; Ettinger et al. 1987; Steineche et al. 1989). Biochemical and histomorphometric data show that the bone protective effect of estrogen in women is associated with decreased bone turnover (Recker et al. 1977; Christiansen et al. 1982; Thomsen et al. 1986; Steineche et al. 1989). Therefore, the skeletal effects of estrogen replacement are similar in estrogen-deplete rats and women.

10.4 Summary

Histomorphometric studies have shown that cancellous bone loss in OVX rats is associated with increased bone turnover. The increment in bone resorption apparently exceeds the increment in bone formation to induce net bone loss. Estrogen replacement suppresses bone turnover and protects against osteopenia in OVX rats. These findings are consistent with the skeletal effects of estrogen depletion and replacement in women.

References

Christiansen C, Christensen MS, Larsen NE, Transbol I (1982) Pathophysiological mechanisms of estrogen effect on bone metabolism. Dose-response relationships in early postmenopausal women. J Clin Endocrinol Metab 55:1124–1130

Elders PJM, Netelenbos JC, Lips P, van Ginkel FC, van der Stelt PJ (1988) Accelerated vertebral bone loss in relation to the menopause. Bone Miner 5:11–19

Ettinger B, Genant HK, Cann CE (1987) Postmenopausal bone loss is prevented by treatment with low-dosage estrogen with calcium. Ann Int Med 106:40–45

Genant HK, Cann CE, Ettinger B, Gordan GS (1982) Quantitative computed tomography of vertebral spongiosa: a sensitive method for detecting early bone loss after oophorectomy. Ann Int Med 97:699–705

Heaney RP, Recker RR, Saville PD (1978) Menopausal changes in bone remodeling. J Lab Clin Med 92:964–970

Horsman A, Simpson M, Kirby PA, and Nordin BEC (1977) Nonlinear bone loss in oophorectomized women. Br J Radiol 50:504–507

Johnston CC, Hui SL, Witt RM, Appledom R, Baker RS, Longcope C (1985) Early menopausal changes in bone mass and sex steroids. J Clin Endocrinol Metab 61:905–911

Kalu DN (1991) The ovariectomized rat model of postmenopausal bone loss. Bone Miner 15:175–192

Kalu DN, Liu CC, Salerno E, Hollis B, Echon R, Ray M (1991) Skeletal response of ovariectomized rats to low and high doses of 17β-estradiol. Bone Miner 14:175–187

Kelly PJ, Pocock NA, Sambrook PN, Eisman JA (1989) Age and menopause-related changes in indices of bone turnover. J Clin Endocrinol Metab 69:1160–1165

Lindsay R, MacLean A, Kraszewski A, Hart DM, Clark AC, Garwood J (1978) Bone response to termination of estrogen treatment. Lancet 2:1325–1327

Nilas L, Borg J, Christiansen C (1984) Different rates of loss of trabecular and cortical bone after the menopause. In: Christiansen C, Arnaud CD, Nordin BEC, Parfitt AM, Peck WA, Riggs BL (eds) Osteoporosis 1. Glostrup Hospital, Copenhagen, pp 457–461

Nilas L, Christiansen C (1988) Rates of bone loss in normal women: evidence of accelerated trabecular bone loss after the menopause. Eur J Clin Invest 18:529–534

Recker RR, Saville PD, Heaney RP (1977) Effect of estrogens and calcium carbonate on bone loss in postmenopausal women. Ann Int Med 87:649–655

Riis BJ, Rodbro P, Christiansen C (1986) The role of serum concentrations of sex steroids and bone turnover in the development and occurrence of postmenopausal osteoporosis. Calcif Tissue Int 38:318–322

Steineche T, Hasling C, Charles P, Eriksen EF, Mosekilde L, Melsen F (1989) A randomized study on the effects of estrogen/gestagen or high dose oral calcium on trabecular bone remodeling in postmenopausal osteoporosis. Bone 10:313–320

Stepan JJ, Pospichal J, Presl J, Pacovsky V (1987) Bone loss and biochemical indices of bone remodeling in surgically-induced postmenopausal women. Bone 8:279–284

Thomsen K, Riis B, Christiansen C (1986) Effect of estrogen/gestagen and 24R,25-dihydroxy vitamin D therapy on bone formation in postmenopausal women. J Bone Miner Res 1:503–507

Turner RT, Evans GL, Wakley GK (1993) Mechanism of action of estrogen on cancellous bone balance in tibiae of ovariectomized growing rats: inhibition of indices of formation and resorption. J Bone Miner Res 8:359–366

Wronski TJ, Cintron M, Doherty AL, Dann LM (1988) Estrogen treatment prevents osteopenia and depresses bone turnover in ovariectomized rats. Endocrinology 123:681–686

Wronski TJ, Dann LM, Scott KS, Cintron M (1989a) Long-term effects of ovariectomy and aging on the rat skeleton. Calcif Tissue Int 41:360–366

Wronski TJ, Dann LM, Horner SL (1989b) Time course of vertebral osteopenia in ovariectomized rats. Bone 10:295–301

Wronski TJ, Dann LM, Scott KS, Crooke LR (1989c) Endocrinology 125:810–816

Wronski TJ, Yen CF, Scott KS (1991) Estrogen and diphosphonate treatment provide long-term protection against osteopenia in ovariectomized rats. J Bone Miner Res 6:387–394

Wronski TJ, Yen CF, Qi H, Dann LM (1993a) Parathyroid hormone is more effective than estrogen or bisphosphonates for restoration of lost bone mass in ovariectomized rats. Endocrinology 132:823–831

Wronski TJ, Dann LM, Qi H, Yen CF (1993b) Skeletal effects of withdrawal of estrogen and diphosphonate treatment in ovariectomized rats. Calcif Tissue Int 53:210–216

11 Biochemical Assessment of Bone Metabolism in Rodents

Roman C. Mühlbauer

11.1 Introduction

This is not an exhaustive review of biochemical methods to assess bone resorption and bone formation in rats and mice but is complementary to the subjects presented by the other contributors to this volume. A number of methods are discussed which have proven to be powerful research tools, mostly in our institution, and have led, for example, to the clinical development of several bisphosphonates. These should also be useful tools for establishing the efficacy of analogs of sex steroids as agents to prevent and possibly cure bone loss.

While it seems that as a preventive treatment, an inhibitor of bone resorption may be sufficient, a curative treatment ideally should be a drug which also increases bone formation. Therefore, reliable animal models are needed which allow both the resorption and formation of bone to be measured.

11.2 Models of Pharmacologically Stimulated Bone Resorption

If rational drug design is not possible, often large numbers of compounds must be screened. For this purpose, reliable, rapid, and potentially simple methods are needed.

In animal models based on the inhibition of pharmacologically stimulated bone resorption by a novel compound, calcaemia has been used as an index of bone resorption. However, the measurement of calcamia can be considered reliable only if its regulation by parathyroid hormone (PTH) and calcitonin is excluded, i.e., the animals are thyroparathyroidectomized, and the flux of calcium from the intestine is negligible. This is achieved either by feeding the rats a low calcium diet, or by overnight fasting before a blood sample is taken. With such methodology, a large number of active bisphosphonates were identified using stimulation of bone resorption by PTH (Fleisch et al. 1969; Russell et al. 1970; Mühlbauer and Fleisch 1981; Benedict et al. 1986; Thompson et al. 1991; Kudo et al. 1992) or $1,25(OH)_2D_3$ (Green et al. 1992). However, in animals treated with PTH, calcemia may not be reliable as an index of bone resorption. Indeed, plasma calcium will be influenced by the stimulatory effect of PTH on the renal reabsorption of calcium. Although bisphosphonates, for example, do not directly influence the renal handling of calcium, interference with this mechanism cannot be excluded for newly developed classes of compounds. Furthermore, it is feared that due to the renal effects of PTH, false negative results may be generated.

Pharmacological means to stimulate bone resorption without the concomitant calcemic effects of PTH have been investigated. A vitamin A analog (ethyl p-[(E)-2-(5,6,7,8-tetrahydro-5,5,8,8-tetramethyl-2-naphthyl)-1-propenyl]benzoate) efficiently stimulated bone resorption but was devoid of a stimulatory effect on renal calcium reabsorption (Trechsel et al. 1987). Its action seems to be mediated by an increased osteoclast number, secondary to an increased proliferation of precursor cells, since irradiation of the rats prevented the hypercalcemia. Bone formation was unchanged as assessed with ^{45}Ca kinetics. Due to these properties, a screening method was developed in which this arotinoid replaced PTH. Because of the high sensitivity and reproducibility of this method, the effect of bisphosphonates on bone re-

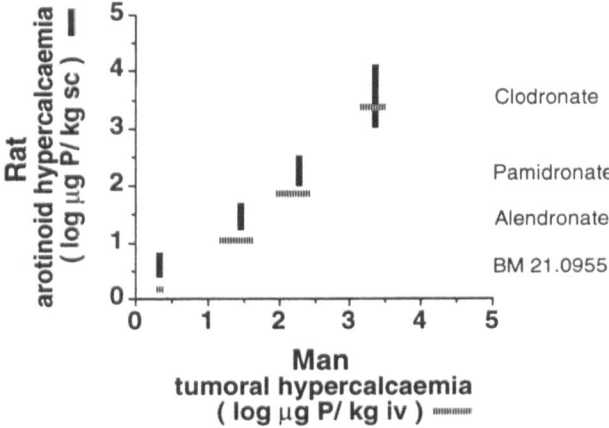

Fig. 1. Comparison of the range of effective doses of various bisphosphonates to inhibit hypercalcemia in rats and man (Copyright R.C. Mühlbauer 1993, used with permission). The results for rats were obtained with the arotinoid test (Mühlbauer et al. 1992). For the sake of comparison with human data, the total dose given during the whole assay period (3 days) was used. Fifty percent inhibition of arotinoid-induced increase of plasma calcium was considered the lowest effective dose. For humans the total dose of bisphosphonates effective (single as well as multiple infusions) to lower calcemia in tumor-induced bone disease, as summarized by Fleisch (1991) were used. The range of doses given as milligram compound were transformed into milligram phosphorus (P) per kilogram body weight, assuming a mean body weight of 70 kg

sorption could be investigated with a much smaller number of animals than previously needed. A large number of compounds have therefore been investigated (Sietsma et al. 1989; Mühlbauer et al. 1992) and the most potent of them selected for further development. This animal model has proved to be very powerful in the detection of inhibitors of bone resorption. The question, however, remains whether the results obtained can be extrapolated to man. Figure 1 compares the potency and range of effective doses of four bisphosphonates inhibiting hypercalcemia in the rat and in man. The correlation found is excellent, indicating the usefulness of the arotinoid model. This good predictive value for doses effective in man, based on body weight may, however, exquisitely depend on the fact that bisphosphonates are not known to be metabolized in vivo. Thus, differences in metabolic rate between the rat and man are probably not relevant in this special situation. For compounds which undergo metabolic transformation, a less impressive correlation may be presumed.

The method has, however, the following limitations: the effect of a compound must occur within 3 days and weak inhibitors of bone resorption may not be detected because of the strong stimulation of bone resorption by the arotinoid. While the first problem may be solved by pretreating the rats with the compound under investigation, the second situation needs an alternative methodology such as described below.

11.3 Continual Monitoring of Endogenous Bone Resorption

The method discussed here, based on the urinary excretion of $[^3H]$tetracycline ($[^3H]Tc$) from prelabeled animals, is not a screening method but is suited for basic research and also for in-depth investigation of selected compounds. The measurement of the urinary excretion allows the monitoring of long-lasting effects of an experimental maneuver on bone resorption as well as effects lasting less than 24 h (see below). Therefore, it is an ideal method to determine the time course of an effect on bone resorption.

Tetracycline is deposited in hard tissues during their formation. This property is widely used in bone histology to determine the rate of bone formation. At a later time, when the labeled bone is resorbed, the tetracycline is freed, circulates in the blood (Wong and Klein 1983; Wong et al. 1983) and is excreted into the urine (Klein and Wong 1985; Mühlbauer and Fleisch 1986) where it can be assessed by measuring the $[^3H]Tc$.

Furthermore, $[^3H]Tc$, once liberated from bone, is only poorly reutilized during bone turnover, as shown in dogs and rabbits (Klein et al. 1985) and in the rat (Mühlbauer and Fleisch 1990a). This is probably due to the fact that $[^3H]Tc$ released from bone circulates in a form which binds poorly to apatite (Mühlbauer and Fleisch 1990a) and to an efficient renal excretion which precludes, for example, ^{45}Ca from being used in a similar way due to its avid tubular reabsorption and reutilization.

Based on these characteristics, a method for continual monitoring of bone resorption has been developed using the urinary excretion of [³H]Tc from chronically prelabeled rats (Mühlbauer and Fleisch 1990a) and mice (König et al. 1988). The urinary excretion of [³H]Tc instead of plasma levels has been chosen according to a basic principle of renal physiology. In steady state the mass of [³H]Tc excreted into urine is independent of the renal handling, i.e., the glomerular filtration rate (GFR) while plasma levels of [³H]Tc are influenced both by the input of [³H]Tc from bone and GFR. According to the same principle, an increase in plasma creatinine is used in clinical medicine to assess a deterioration of GFR, while the mass of creatinine excreted into the urine does not decrease until very low values of GFR are reached and reflects the input of creatinine into plasma from muscle and diet. This is due to the compensatory increase in plasma concentration which, in spite of a decrease in GFR, maintains the filtered load constant. This principle has recently been challenged, and it has been proposed that serum, rather than urine, measurements of [³H]Tc would be preferable (Golomb et al. 1992). These authors found, however, that using a dose of 160 μmol/kg per day for 6 days of the bisphosphonate pamidronate, which was substantially above the maximum dose needed to inhibit bone resorption (Reitsma et al. 1980; Mühlbauer et al. 1992), and which was nephrotoxic as shown by an increase in serum creatinine levels, did not produce a significant decrease in serum [³H]Tc as expected according to the principle described above. Interestingly, at a four times lower dose which, however, did not lead to an increase in plasma creatinine, a significant inhibition of bone resorption was detected. These results confirm that indeed blood levels are exquisitely sensitive to changes in GFR, which makes results from serum [³H]Tc levels difficult to interpret.

Using 24-h urine collections in rats, a strong stimulation of endogenous bone resorption has been found with dietary calcium restriction, PTH, or arotinoid administration. Dietary calcium supplementation, thyroparathyroidectomy, and bisphosphonate administration inhibited bone resorption (Mühlbauer and Fleisch 1990a; Mühlbauer et al. 1991). Especially relevant to the topic of this workshop is that a significant increase in bone resorption has also been found in less than 1 week after ovariectomy (Mühlbauer and Fleisch 1990b) using this method. Finally, using a modification of the method, the inhibitory effect of calcitonin on bone resorption has also been shown (Sjögren et al. 1991). Except in the latter study, all these effects must occur in "old bone", since the time elapsing between the last [³H]Tc injection for labeling (two injections per week for the first 6 weeks of life) and the experimental maneuver is about 20 days (10 days equilibration, 10 days baseline urine collection). Since in the case of the primary spongiosa of proximal tibiae in growing rats, the life span of this tissue is about 5 days (Kimmel and Jee 1980) and [³H]Tc reincorporation negligible, the effects observed must stem from secondary spongiosa and compact bone, i.e., "old bone".

In addition, as dietary calcium intake has been shown to have a strong effect in this system (Mühlbauer and Fleisch 1990a), the basal level of bone resorption can be selected by the calcium content of the diet used. Thus, the sensitivity to inhibitors of bone resorption may be adjusted accordingly.

If urine from [³H]Tc prelabeled rats is collected at intervals shorter than 24 h, such as, every 6 h, this method also allows the detection of effects of short duration, for example, a strong diurnal rhythm of bone resorption has been established using this technique (Mühlbauer and Fleisch 1990a). Peak [³H]Tc values were found in the 6-h urine specimens following food intake. This diurnal rhythm appears to be independent of PTH and calcitonin since it persists in thyroparathyroidectomized rats. When the daily amount of food and water was given in four equal portions every 6 h, the large increase in bone resorption occurring after a single food administration was blunted and overall bone resorption decreased conspicuously. Half of this effect was due to altering the intake habit of solid food and the other half by altering the intake of water. These results were confirmed by a calcium balance and, when the "nibbling" was prolonged for 30 days, also led to a significant increase of trabecular bone volume as well as whole body calcium (Mühlbauer and Fleisch 1992). Furthermore, when bone resorption was pharmacologically inhibited with a bisphosphonate, the postprandial peak of bone resorption expressed as percentage of the daily resorption was unchanged (Mühlbauer and Fleisch 1990a). This suggests that an adequate modulation of bone resorption for calcium homeostasis still persists under treatment with a potent inhibitor of bone resorption, as a bisphosphonate.

The body weight of the rats used in these experiments was up to 300 g. If only small amounts of compounds are available or the price for the treatment of animals this size is prohibitive, this problem can be solved by adapting the [^{3}H]Tc method to the mouse (König et al. 1988). This animal is especially well suited for the investigation of the effects of cytokines on bone resorption, since normally only murine or human cytokines are available. Indeed with this technique, a strong stimulation of bone resorption following systemic administration of interleukin-1α and tumor necrosis factorα was shown (König et al. 1988). Due to the small urine volumes, which require daily rinsing of the metabolic cages, the use of the [^{3}H]Tc technique in the mouse is more cumbersome than in the rat. The sensitivity and precision of the murine [^{3}H]Tc technique is, however, not inferior to the one observed in the rat as judged by the response to reduced calcium intake, infusion of PTH, dietary calcium supplementation, or bisphosphonate administration (König et al. 1988). However, some differences between species were observed: the mouse appears resistant to the effect of the arotinoid which efficiently stimulates bone resorption in the rat (Mühlbauer and Fleisch, unpublished observation), and no stimulation of bone resorption by ovariectomy was observed in the mouse (König et al., unpublished observation).

11.4 Assessment of Bone Mass

The ultimate objective for a drug used in the treatment of osteoporosis is to increase bone mass. Techniques which need the sacrifice of small laboratory animals to determine the amount of bone by calcium or ash weight, histology, or single photon absorptiometry on explanted bones have the inherent methodological limitations that, due to interanimal variation, differences in bone mass between groups have to be relatively large or the number of animals exorbitant in order to reach statistical significance.

Recently, it has been shown that dual-energy X-ray is an accurate and precise method, also in the rat, for the noninvasive measurement of total bone mineral content (Casez et al. 1992). The precision and sensitivity of this method is such that the mineral density of vertebrae and individual long bones can also be followed (Ammann et al. 1992).

With this technique, significant differences at the level of the lumbar spine and the proximal tibia, both regions with a high proportion of trabecular bone, were found 30 days after ovariectomy. In more distal regions of the tibia, consisting essentially of cortical bone, no significant differences were seen, confirming previous reports in which bone mass in ovariectomized rats was assessed by histological means (Wronski et al. 1988, 1989). This technique is, therefore, promising for use in future developments where bone mass needs to be followed up.

The sensitivity of this technique to assess bone mass is such that sufficient cumulation of the effect is required in order to reach statistical significance. Therefore, the effect of a treatment lasting less than 1 month may not be detected. Since calcium retained is retained virtually exclusively in bone, whole body net calcium retention may therefore be used to assess short term effects on bone mass.

11.5 Calcium Balance Studies

From the difference of calcium ingested and the excretion into feces and urine, the net whole body calcium retention can be measured over 3 days with a reasonable precision and repetitive measurements are also possible. To avoid spurious results, however, a prior equilibration of the rats to the experimental conditions of at least 1 week is necessary. Using this technique, an increased daily calcium retention was found in rats treated with bisphosphonates (Gasser et al. 1972; Reitsma et al. 1980). This result is consistent with the accumulation of bone mass in growing rats treated with these compounds. It is, however, not possible with balance studies to find out whether an effect is due to a change of resorption or formation of bone or both.

11.6 ^{45}Ca Kinetics

From a balance study of stable calcium and the disappearance curve of intravenously injected ^{45}Ca in the same animal, many variables of calcium metabolism can be obtained by analyzing the data according to a two-compartment model (Richelle 1967; Gasser et al. 1972; Morgan et al. 1975; Bonjour et al. 1975; Haldimann et al. 1977). Among these, the flux of calcium into bone (bone formation) and out of bone (bone resorption) are the most relevant. ^{45}Ca kinetic studies represent one of the most complete methods of investigation of calcium metabolism in the rat. Since steady-state conditions are required, the rats must be equilibrated to the experimental conditions as well as to the treatment for at least 1 week. Furthermore, with this technique the disappearance of ^{45}Ca from the rapidly exchangeable pool that is not accounted for by urinary and endogenous fecal excretions is used to assess the entry of calcium into bone. Therefore, the experiment must be of short duration (3 days) and cannot be repeated in the same animal in order to avoid a bias by the subsequent exit of ^{45}Ca from bone. Using this method, it has been found, for example, that a low dose (1 mg P/kg) of the bisphosphonate etidronate significantly inhibits bone resorption with little effect on bone formation (Gasser et al. 1972). At a tenfold higher dose, however, bone resorption as well as the flux of calcium into bone decreases dramatically. The decrease is associated with an increased urinary output as well as a reduced intestinal absorption of calcium, leading to a marked decrement in calcium retention. This high dose of etidronate inhibits mineralization (Fleisch et al. 1971) and decreases $1,25(OH)_2D_3$ synthesis. Treatment with this vitamin D metabolite prevents the fall in intestinal calcium absorption but fails to restore the defect in mineralization (Bonjour et al. 1973,1975), indicating that etidronate has a direct effect on this process.

11.7 Conclusion

The techniques discussed in this chapter have been of invaluable help in developing potent inhibitors of bone resorption, i.e., the bisphosphonates. A large body of knowledge has been obtained from young rats in which the growth-dependent modeling drifts are overwhelming and the bone multicellular unit based turnover not appreciable. Consequently, it could be feared that results generated in young rats are of limited relevance in adult humans. Nevertheless, the information from these models was not misleading with respect to the inhibition of bone resorption in man. Indeed it appears that in growing rats, the effect of a potential inhibitor of bone resorption can be assessed in a limited length of time since it is amplified by the inhibition of the rapid growth dependent modeling. Therefore, these animal models represent a strategy with optimal cost-benefit ratio to determine the efficacy of an inhibitor of bone resorption.

Together with other techniques, such as bone histomorphometry, which allows the specific site of effect to be determined, these efficient rodent models represent a promising strategy for future research on drugs affecting bone metabolism.

Acknowledgments. I wish to thank Dr. M.G. Cecchini for critical review. Many thanks also go to I. Ryf for typing the manuscript and to G. Mühlbauer for retrieving the literature, to Dr. C.M. Lim-Taylor for correcting the English typescript, and to O. Aeby for the help with the illustration.

References

Ammann P, Rizzoli R, Slosman D, Bonjour JP (1992) Sequential and precise in vivo measurement of bone mineral density in rats using dual-energy X-ray absorptiometry. J Bone Miner Res 7/3:311–316

Benedict JJ, Johnson KY, Bevan JA, Perkins CM (1986) A structure/activity study of nitrogen heterocycle containing bis(phosphonates) as bone resorption inhibiting agents. Calcif Tissue Int 38 [Suppl]: S31 (abstract 114)

Bonjour JP, DeLuca HF, Fleisch H, et al (1973) Reversal of the EHDP inhibition of calcium absorption by 1,25-dihydroxycholecalciferol. Eur J Clin Invest 3:44–48

Bonjour JP, Trechsel U, Fleisch H, Schenk R, DeLuca HF, Baxter LA (1975) Action of 1,25-dihydroxyvitamin D_3 and a diphosphonate on calcium metabolism in rats. Am J Physiol 229:402–408

Casez JP, Mühlbauer RC, Lippuner K, et al (1992) Dual-energy X-ray absorptiometry (DXA) is an accurate and precise method for measuring rat total bone mineral content. J Bone Miner Res 7 [Suppl 1]:S267 (abstract 697)

Fleisch H (1991) Bisphosphonates: Pharmacology and use in the treatment of tumour-induced hypercalcaemic and metastatic bone disease. Drugs 42:919–944

Fleisch H, Russell RGG, Francis MD (1969) Diphosphonates inhibit hydroxyapatite dissolution in vitro and bone resorption in tissue culture and in vivo. Science 165:1262–1264

Fleisch H, Bisaz S, Schenk R, Russell RGG (1971) Pyrophosphate, pyrophosphatases, diphosphonates and mineralisation. In: Metabolism. Proceedings of the 13th International Congress of Pediatrics, Vienna, August 29 – September 4, pp 265–270

Gasser AB, Morgan DB, Fleisch HA, Richelle LJ (1972) The influence of two diphosphonates on calcium metabolism in the rat. Clin Sci 43:31–45

Golomb G, Eitan Y, Hoffman A (1992) Measurement of serum [^{3}H]tetracycline kinetics and indices of kidney function facilitate study of the activity and toxic effects of bisphosphonates in bone resorption. Pharm Res 9/8:1018–1023

Green JR, Müller K, Jaeggi KA (1992) Pharmacological characterization of bisphosphonate compounds containing a basic nitrogen substituent. Bone Miner 17 [Suppl 1]:S12 (abstract 6)

Haldimann B, Bonjour JP, Fleisch H (1977) Role of parathyroid hormone in regulation of main calcium fluxes in rats. Am J Physiol 232:E535-E541

Kimmel DB, Jee WSS (1980) Bone cell kinetics during longitudinal bone growth in the rat. Calcif Tissue Int 32:123–133

Klein L, Wong KM (1985) Rapid effect of calcium deficiency upon bone resorption in rats prelabeled with [^{3}H]tetracycline and ^{45}Ca. Clin Res 33:889A (abstract)

Klein L, Wong KM and Simmelink JW (1985) Biochemical and autoradiographic evaluation of bone turnover in prelabeled dogs and rabbits on normal and calcium-deficient diets. Bone 6:395–399

König A, Mühlbauer RC, Fleisch H (1988) Tumor necrosis factor and interleukin-1 stimulate bone resorption in vivo as measured by urinary [^{3}H]tetracycline excretion from prelabeled mice. J Bone Miner Res 3:621–627

Kudo M, Abe T, Motoie H, et al (1992) Pharmacological profile of new bisphosphonate, 1-hydroyy-2-(imidazo[1, 2-a]pyridin-3-yl)ethane-1,1-bis(phosphonic acid). Bone Miner 17 [Suppl 1]:S13 (abstract 11)

Morgan DB, Gasser A, Largiadès U, Jung A, Fleisch H (1975) Effects of a diphosphonate on calcium metabolism in calcium-deprived rats. Am J Physiol 228:1750–1756

Mühlbauer RC, Bauss F, Schenk R, et al (1991) BM 21.0955, a potent new bisphosphonate to inhibit bone resorption. J Bone Miner Res 6:1003–1011

Mühlbauer RC, Fleisch H (1981) Effect of various polyphosphates on ectopic calcification and bone resorption in rats. Miner Electrolyte Metab 5:296–303

Mühlbauer RC, Fleisch H (1986) Bone resorption monitored with urinary excretion of ^{3}H-tetracycline from prelabelled rats. In: Delmas PD, Mennier P, Cohn DU (eds) Program and Abstracts IXth International Conference on Calcium Regulating Hormones and Bone Metabolism, p 295

Mühlbauer RC, Fleisch H (1990a) A method for continual monitoring of bone resorption in rats: evidence for a diurnal rhythm. Am J Physiol 259:R679-R689

Mühlbauer RC, Fleisch H (1990b) Dietary administration of fish oil does not inhibit bone resorption in intact and ovariectomized rats. Calcif Tissue Int 46 [Suppl 2]:A25 (abstract 94)

Mühlbauer RC, Fleisch H (1992) Inhibition of bone resorption and increase of bone mass by feeding rats 4 times daily instead of once. Bone Miner 17 [Suppl 1]:92 (abstract 79)

Reitsma PH, Bijvoet OLM, Verlinden-Ooms H, van der Wee-Pals LJA (1980) Kinetic studies of bone and mineral metabolism during treatment with (3-amino-1-hydroxypropylidene)-1,1-bisphosphonate (APD) in rats. Calcif Tissue Int 32:145–157

Richelle LJ (1967) Contribution à l'étude du métabolisme minéral de l'os chez le rat. Thesis, University of Liège

Russell RGG, Mühlbauer RC, Bisaz S, Williams DA, Fleisch H (1970) The influence of pyrophosphate, condensed phosphates, phosphonates and other phosphate compounds on the dissolution of hydroxyapatite in vitro and on bone resorption induced by parathyroid hormone in tissue culture and in thyroparathyroidectomised rats. Calcif Tissue Res 6:183–196

Sietsema WK, Ebetino FH, Salvagno AM, Bevan JA (1989) Antiresorptive dose-response relationships across three generations of bisphosphonates. Drugs Exp Clin Res 15:389–396

Sjögren S, Rydström K, Wallmark B (1991) The effects of clodronate and salmon calcitonin on daily rat bone resorption. J Bone Min Res 6 [Suppl 2]:S175 (abstract 368)

Thompson DD, Seedor JG, Grasser W, Rosenblatt M, Rodan GA (1991) Effect of alendronate (bisphosphonate) in animal models of hyperparathyroidism. Contrib Nephrol 91:134–139

Trechsel U, Stutzer A, Fleisch H (1987) Hypercalcemia induced with an arotinoid in thyroparathyroidectomized rats. A new model to study bone resorption in vivo. J Clin Invest 80:1679–1686

Wong KM, Klein L (1983) Effects of varying doses of 3 diphosphonates (EHDP, Cl$_2$MDP, & APD) on bone resorption and formation in rats prelabeled with ^{3}H-tetracycline. Program and abstracts of the 5th annual scientific meeting of the American Society for Bone and Mineral Research, San Antonio, TX (Abstract) A17

Wong KM, Klein L, Hollis BW (1983) Diurnal assessment of plasma 1,25(OH)$_2$D, 25(OH)D, [45]Ca and ^{3}H-tetracycline (bone resorption) in prelabelled dogs. Program and abstracts of the 5th annual scientific meeting of the American Society for Bone and Mineral Research, San Antonio, TX (Abstract) A69

Wronski TJ, Cintron M, Dann LM (1988) Temporal relationship between bone loss and increased bone turnover in ovariectomized rats. Calcif Tissue Int 43:179–183

Wronski TJ, Dann LM, Horner SL (1989) Time course of vertebral osteopenia in ovariectomized rats. Bone 10:295–301

Subject Index